STAGE FRIGHT

Theory of the Partisan

Theory of the Partisan

Intermediate Commentary on the Concept of the Political

Carl Schmitt

Translated by G. L. Ulmen

Telos Press Publishing
New York

Printed in the United States of America
17 16 15 14 13 12 3 4 5 6

Originally published in German as *Theorie des Partisanen: Zwischenbemerkung zum Begriff des Politischen*, 2nd ed.,

ISBN: 978-0-914386-33-9

Library of Congress Cataloging-in-Publication Data

Schmitt, Carl, 1888–1985.
[Theorie des Partisanen English]
Theory of the partisan : intermediate commentary on the concept of the political / Carl Schmitt ; translated by G. L. Ulmen.
p. cm.
ISBN 978-0-914386-33-9 (pbk. : alk. paper)
1. Guerrilla warfare. I. Title.
U240.S311 2007
355.02'18—dc22

2007018126

Telos Press Publishing
431 East 12th Street
New York, NY 10009

www.telospress.com

Dedicated to Ernst Fortshoff
on his 60th Birthday
September 13, 1962

Contents

Translator's Introduction*

The origin of the term "partisan" is uncertain, and what precisely constitutes "partisan warfare" is a matter of some dispute. Generally speaking, the term "partisan" designates either the leader or a member of "light troops" or participants in a "small war." So-called "irregular warfare" by regular as well as irregular troops and even armed civilians has been known in Europe since the 18th century. Partisan warfare was notable in the Seven Years War (1754 and 1756–1763), more specifically in the theater of the French and Indian War in America. In the American Revolutionary War, George Washington was able to defeat a much larger English army during the so-called "forage war" by utilizing partisan warfare in New Jersey, even though he abhorred the idea.[1] By 1780, Francis Marion, the so-called "Swamp Fox," had seen enough of war to realize that the Continentals were overlooking a very useful technique—partisan warfare. He obtained permission to organize a company that at first consisted of twenty ill-equipped men and boys, and his guerrilla activities in South Carolina soon began to toll heavily on the British, especially Cornwallis.

* The translation of *Theory of the Partisan* that appeared in *Telos*, No. 127 (Spring 2004) is unreliable. Changes were made to my translation that I refused to accept, which is why I removed my name, and why the present translation is being published.

1. See David Hackett Fischer, *Washington's Crossing* (New York: Oxford University Press, 2004), pp. 346ff.

Johann Ewald, who had extensive experience as a company commander in the Hessian Field Jaeger Corps during both the French and Indian War and the American Revolutionary War, published a treatise on "small war" in 1785, which has been translated as "partisan warfare."[2] The translators acknowledge that some readers may object, since it might appear that they are applying a "modern term" to premodern events, given that "partisan," especially in the 20th century, implied a *political* component. But they rightly point out that several conflicts in early modern Europe as well as in the American Revolutionary War contained "an ideological component" that distinguished them from regular warfare.[3] Indeed, they offer a number of examples from both 18th and 19th century sources attesting to use of the term "partisan" and emphasizing the *political* character of the term. One of the most telling and significant with respect to Schmitt's treatise is the definition *Parthey, Parti* in Johan Heinrich Zedler's 1740 dictionary: "a group of soldiers on horseback or on foot, which is sent out by a general to do damage to the enemy by ruses and speed, or to investigate his condition.... It has to have valid passports, letters of marque or salviguards, otherwise they are considered highway robbers. The leader of such a party is called a *Partheygänger* [party follower] or partisan."[4]

The question then arises as to how earlier treatises recognizing a distinct *political* component relate to Schmitt's "theory" of the partisan. Perhaps the obvious point of comparison is that all these earlier treatises, including Ewald's, were more descriptive than theoretical, more military than political. But even here, the distinction is imprecise. In their introductory essay, the

2. See Johann Ewald, *Treatise on Partisan Warfare*, tr. Robert A. Selig and David Curtis Skaggs (New York, Westport, CN, and London: Greenwood Press, 1991). Cf. Johann von Ewald, *Diary of the American Revolutionary War: A Hessian Journal* (New Haven: Yale University Press, 1979).

3. Ibid., p. 5.

4. *Grosses Vollständiges Universal Lexicon* (Leipzig and Halle: J. H. Zedler, 1732–1750), Vol. 26, p. 1050.

translators of Ewald's treatise write: "What enabled the colonists to defend successfully their political experiment by means of armed resistance was the genuine fusion of military tactics and political motivation. The colonists diluted the boundaries between regular light troops (permanent members of the field army) and militiamen (temporarily armed civilians) as carriers of irregular warfare."[5] At another point, they observe the limitations of Ewald as a theoretician, because he constantly moves "back and forth between regular light infantry warfare and what we would call guerrilla warfare."[6] Here then, a distinction is drawn between regular light troops and militiamen, and between partisan warfare and guerrilla warfare, which helps to focus on the more precise meaning of the term "partisan." Finally, they observe that if Ewald's seven years taught him anything, it was surely that "the 'well trained and well-disciplined soldiers' of Great Britain and her German allies had been defeated by a combination of irregular and regular forces that they had never before encountered in Europe."[7] This statement supports Schmitt's assertion that the first example of guerrilla operations on a grand scale was in Spain between 1808 and 1813, and that is why Schmitt begins his historical and theoretical treatise at this point.

Mao Tse-tung, certainly the greatest theoretician of partisan warfare, also recognized the Spanish case to be a watershed.[8] In fact, the term *guerrilla*, which literally means "small war," stems from the Spanish national struggle against the French,[9] although in the 20th century the term "guerrilla" usually was reserved for internal opponents of a government or occupying force, such as

5. Ewald, *Treatise on Partisan Warfare*, op. cit., p. 24.

6. Ibid., p. 26.

7. Ibid., p. 27.

8. See Mao Tse-tung, *On Guerrilla Warfare*, tr. Samuel B. Griffith II (Urbana and Chicago: University of Illinois Press, 2000), p. 10.

9. Ian F. W. Beckett, *Encyclopedia of Guerrilla Warfare* (New York: Checkmark Books, 2001), p. xi.

the Viet Cong during the Vietnam War or the PLO in Palestine. Mao is certainly right in asserting that only when Lenin came on the scene did "partisan warfare" receive the potent *political* injection that was to alter its character radically. But Mao insisted that guerrilla operations must not be considered an independent form of warfare, but rather one step in the total war, one aspect of the revolutionary struggle.[10] During the American Civil War (1861–1865), there were Confederate partisan fighters who were nothing more than terrorists. However, whereas the distinction between "light troops," "irregular troops," "guerillas," and "partisans" is not always clear, the distinction between "regular troops," i.e., soldiers, "partisans," and "terrorists" is both conceptually and politically clear.

Schmitt sticks to what he calls the "classical" definition of the partisan, which means that the term refers to specific historical figures and situations. Conceptually, as Schmitt says, the horizon of his investigation ranges from the Spanish guerrilla of Napoleonic times to the well-equipped partisan of the 1960s, from Empecinado, via Mao Tse-tung and Ho Chi-minh, to Fidel Castro and Raul Salan. Of course, not everything that Schmitt has to say about partisans is original, any more than everything Ewald wrote in his treatise was new. However, by avoiding loose definitions, Schmitt is able to identify specific characteristics of the partisan and to place him in the historical context determined by the collapse of the European state system.[11] For one, the partisan is an *irregular* fighter, which means that his actual target is the *regular* soldier in uniform. Despite this distinction, the armed partisan remains dependent on cooperation with a regular organization; he cannot exist in a political no-man's-land. In addition to irregularity, another characteristic of the partisan is his increased mobility of combat. Then, there is his *telluric*

10. Mao, *On Guerrilla Warfare*, op. cit., p. 41.

11. See Carl Schmitt, *The* Nomos *of the Earth in the International Law of the* Jus Publicum Europaeum, tr. G. L. Ulmen (New York: Telos Press, 2003).

character, i.e., his basically *defensive* posture, meaning that his hostility is spatially limited. As Schmitt was well aware, "a *motorized* partisan loses his *telluric* character and becomes only the transportable and exchangeable tool of a powerful central agency of world politics, which deploys him in overt or covert war, and deactivates him as the situation demands." This *motorized* partisan already was a sign of a changed historical situation, even as is Schmitt's observation that: "As regards contemporary partisans, the antithetical pairs of regular-irregular and legal-illegal usually become blurred and interchangeable." Nevertheless, until today the partisan remains an *irregular* fighter, and combat against partisans is often only a mirror image of partisan warfare.

Schmitt also observes that from World War II until today, despite all characteristic links and mixtures, there are still basically only two types of partisans: defensive, authochtonous defenders of the homeland; and globally aggressive revolutionary activists. The antithesis between the two remains, because each is based on a fundamentally different type of war and enmity. Even though the various types of partisan warfare still are confused and fused in the praxis of contemporary warfare, they remain so different in their fundamental presuppositions that they "test the criterion of the friend-enemy opposition."[12] As Schmitt notes, the 1949 Geneva Conventions were already an indication of the disintegration of traditional European international law based on the state, even though the uniform codes have only European experiences in view, i.e., not Mao Tse-tung's partisan warfare. Moreover, the degree to which the essential distinctions of "classical" positions were loosened or even challenged opened the door "for a type of war that consciously destroys these clear separations. Then, many discretely stylized compromise norms appear only as the narrow bridge over an abyss, which conceals

12. Cf. Carl Schmitt, *The Concept of the Political*, tr. George Schwab, with Leo Strauss's Notes on Schmitt's Essay, tr. Harvey Lomax (Chicago and London: The University of Chicago Press, 1996).

a successive transformation of concepts of war, enemy, and partisan." After 1945, belligerent acts around the world assumed a distinctly partisan character.

This new situation was brought about by resistance movements against the Nazis in Europe and by the revolutionary program of international communism in Europe and in Asia. Generally speaking, the European movements hewed to the *telluric*-defensive type of partisan warfare, whereas the Asian movements hewed to the revolutionary-aggressive type. After World War II, it was the latter type of partisan warfare that became dominant. Even for Clausewitz, guerrilla warfare was a highly *political* matter, meaning that it had a revolutionary character.[13] What Lenin learned from Clausewitz was not only that war is the continuation of politics by other means, but that the friend-enemy distinction is primary in the age of revolution. Lenin's alliance of philosophy and the partisan, says Schmitt, caused nothing less than the destruction of the whole Eurocentric world that Napoleon had hoped to rescue and that the Congress of Vienna had hoped to restore. For Lenin, partisan warfare was consistent with the methods of civil war.

Generally speaking, civil war is military conflict between two or more approximately equal political governments for sovereignty over people and territory native to both, whereas revolution is a change, not necessarily by force or violence, whereby one system of legality is terminated and another is constituted within the same country.[14] Civil war is horizontal,

13. Cf. Carl von Clausewitz, "Vorlesungen über den Kleinen Krieg," in Clausewitz, *Schriften, Aufsätze, Studien, Briefe*, ed. Werner Hahlweg (Göttingen: Vandenhoek und Ruprecht, 1966). Cf. also Carl Schmitt, "Clausewitz als politischer Denker. Bemerkungen und Hinweise," in Günter Dill, ed., *Clausewitz in Perspektive: Materialien zu Carl von Clausewitz: Vom Kriege* (Frankfurt a/M, Berlin, and Vienna: Ullstein, 1980), pp. 419–46.

14. Cf. Lyford Patterson Edwards, "Civil War," in *Encyclopedia of the Social Sciences*, ed. Edwin R. A. Seligman and Alvin Johnson (New York: The MacMillan Company, 1942), Vol. 3, pp. 523–25.

whereas revolution is vertical. Civil war, as compared with foreign war, is war for complete conquest—the extinction of the enemy government—for at least one side, whereas foreign war, at least in modern times, generally is waged only for certain specific and limited aims. As Schmitt observed, "In civil war, the enemy no longer has any common concepts, and every concept becomes an encroachment on the enemy camp."[15] Indeed: "Civil war has something gruesome about it. It is fraternal war, because it is pursued within a common political unity that includes also the opponent, and within the same legal order, and because both belligerent sides absolutely and simultaneously affirm and negate this common unity. Both consider their opponent to be absolutely and unconditionally wrong. Both reject the right of the opponent, but in the name of the law. Civil war is subject essentially to the jurisdiction of the enemy. Thus, civil war has a narrow, specifically dialectical relation to law. It cannot be anything other than *just* in the sense of being self-righteous, and on this basis becomes the prototype of just and self-righteous war."[16]

Elsewhere, Schmitt wrote: "Just war, i.e., the deprivation of the rights of the opponent in war and the self-improvement of the just side means: transformation of state war (i.e., of war in international law) into a war that is simultaneously colonial war and civil war; that is logical and irresistible; war becomes global civil war and ceases to be war between states."[17] Schmitt further clarified the distinction: "Civil war is *bellum*, with just cause on both sides, but both *sides* in—*justus hostis*. Law, i.e., the form of just war, sanctifies death in war; but how does the reality of civil

15. Carl Schmitt, *Glossarium: Aufzeichnungen der Jahre 1947–1951*, ed. Eberhard Freiherr von Medem (Berlin: Duncker & Humblot, 1991), dated October 31, 1947, p. 36.

16. Carl Schmitt, *Ex Captivitate Salus: Erfahrungen der Zeit 1945/47* (Cologne: Greven Verlag, 1950), pp. 56, 57.

17. Schmitt, *Glossarium*, op. cit., dated October 8, 1947, p. 29.

war or social war compensate for the crime of suicide? Formal legal war justifies killing; the form, not the *justa causa*. That is lacking in civil war. Civil war is not simply war without form (on the contrary, it serves the forms of law and justice), but it's the most gruesome formal destruction of form. Whoever is compelled to participate in civil war is compelled to kill, which is not justified by any formally justified order. Given that he kills, is he obliged to kill, is he obliged to kill because of his opponent? Or is it the total situation of civil war (the self-destruction, the suicide of the *macros anthropos* [large man, i.e., the state]) because the suicide of the individual participant certainly is not justified, also not excused, but only absorbed, and certainly by the specific reality of the situation?"[18]

For Schmitt, the overcoming of civil war was the core of interstate war in the *jus publicum Europaeum*, and during the Eurocentric epoch of bracketed war, partisan warfare remained a marginal phenomenon. But the situation had changed significantly by the early 1960s. Schmitt rightly observed that linking homeland resistance with the aggressiveness of the international communist revolution was dominating partisan warfare worldwide. This is why he insisted that the theory of the partisan was the key to knowledge of political reality. Today, it might be said that the theory of the terrorist is the key to knowledge of political reality. Schmitt's theory of the partisan contains an implicit theory of the terrorist.

Like partisan warfare, terrorism has been around since the beginning of recorded history, meaning that guerrillas throughout history often have resorted to terrorist tactics to achieve their goals. However, terrorism and partisan warfare are not synonymous, and the distinction should be kept in mind. Terrorism emerged in its own right as a separate phenomenon alongside urban guerrilla warfare in the late 1960s and early

18. Schmitt, *Glossarium*, op. cit., dated October 10, 1947, pp. 31–32.

1970s.[19] Like partisan warfare, terrorism is a political concept.[20] It is violence or the threat of violence used in pursuit of a political end. But unlike partisan warfare, terrorism is a planned, calculated, and systematic act. Of course, the definition of terrorism has changed many times since the French Revolution, but after World War II the term regained the revolutionary connotations with which it is most commonly associated today. Like the partisan, the terrorist is irregular; but unlike the partisan, he does not depend on cooperation with a regular base; he can and does exist in a political no-man's-land. Like the partisan, the terrorist is illegal, but unlike the partisan, his illegality is illegitimate, meaning that it has no point of reference. Like the partisan, the terrorist has increased mobility of active combat. But unlike the partisan, he is not essentially defensive; rather, he is essentially *offensive*. Moreover, unlike the partisan, the terrorist is not *telluric*, meaning that his hostility is not spatially limited.[21]

Generally speaking, the partisan has a *real* enemy, who is fought in a war, however irregular, and has some claim to humanity and legitimacy, whereas the terrorist has an *absolute* enemy, who must be annihilated. Here is where the transition from the partisan to the terrorist is most evident, and where the distinction is conceptually clear. In Che Guevara's book on guerrilla warfare, written in 1960, he asserted that it was necessary to distinguish clearly between "sabotage, a revolutionary and highly effective method of warfare, and terrorism, a measure that is generally ineffective and indiscriminate in its results, since it often makes victims of innocent people and destroys a large number of lives

19. Ian F. W. Beckett, *Encyclopedia of Guerrilla Warfare* (New York: Checkmark Books, 2001), p. xiv.

20. Bruce Hoffman, *Inside Terrorism* (New York: Columbia University Press, 1998), p. 14.

21. Cf. J. B. S. Hardman, "Terrorism," in *Encyclopedia of the Social Sciences*, op. cit., Vol. 13, pp. 575–79.

that would be valuable to the revolution."[22] However, in his 1967 "Message to the Tricontinental," Che Guevara made the transition from guerrilla fighter to terrorist. Specifically, he said that hatred is an essential element of the struggle, namely "a relentless hatred of the enemy, impelling us over and beyond the natural limitations that man is heir to and transforming him into an effective, violent, selective, and cold killing machine. Our soldiers must be thus; a people without hatred cannot vanquish a brutal enemy. We must carry the war into every corner the enemy happens to carry it: to his home, to his centers of entertainment; a total war."[23]

For the religious terrorist, violence is first and foremost a sacramental act or divine duty executed in direct response to some theological demand or imperative. Thus, religious terrorism is more theological than political, since its perpetrators are unconstrained by political, moral, or practical constraints that may affect other terrorists. Like Che Guevara in his "Message to the Tricontinental," religious terrorists are engaged in what they regard to be a total war. They seek to appeal to no other constituency than themselves. Their goal is not war as usually understood, but annihilation of the foe. Here Schmitt: "Annihilation of the foe, however, is the attempt (claim) to *creation ex nihilo* [create from nothing], to create a new world on the basis of a *tabula rasa*. Whoever wants to annihilate me is not my enemy, but my satanic pursuer. The question that I should ask in relation to him no longer can be answered politically, but only theologically. The most concrete type of dialectical theology comes about when the annihilator claims that he wants nothing more than to annihilate the annihilator."[24]

22. Che Guevara, *Guerrilla Warfare* (Lincoln: University of Nebraska Press, 1998), p. 21.

23. Ibid., p. 173.

24. Schmitt, *Glossarium*, op. cit., dated June 1, 1948, p. 190.

While political terrorists tend to cloak themselves in the terminology of military jargon, and consciously portray themselves as bona fide fighters, if not soldiers, the most radical religious terrorists do not even demand to be treated as prisoners of war. And they should not be. Terrorists are not soldiers! This is as true for secular terrorists as for religious terrorists. In normal war, i.e., in war as it has been fought during the European epoch of international law, there are rules and accepted norms of behavior that prohibit the use of certain types of weapons, proscribe various tactics, and outlaw attacks on specific categories as targets. In theory, if not always in practice, "the rules of war as codified in the Hague and Geneva conventions, not only grant civilian non-combatants immunity from attack, but also prohibit taking civilians as hostages, impose regulations governing the treatment of POWs, outlaw reprisals against civilians or POWs, recognize neutral territory and the rights of citizens of neutral states."[25] Fighters who do not wear uniforms, have no rank, do not display weapons openly, target civilians, and in some cases commit suicide for mass murder have no status in any conventions of war. One of the fundamental principles, if one can call it that, of international terrorism is a refusal to be bound by rules of warfare and codes of conduct.[26] Even a cursory view of terrorist practices over the past few years demonstrates that they have violated all such rules and codes. In Schmitt's terms, terrorists are "outside the law."

What does this mean concretely? Outside what law? It means outside international law in general and the laws of war

25. Hoffman, *Inside Terrorism*, op. cit., p. 34.

26. It is significant to note that both the Hague and the Geneva conventions occurred after the collapse of the *jus publicum Europaeum*. See *The Hague Peace Conferences of 1899 and 1907 and International Arbitration. Reports and Documents*, ed. Shabtai Rosenne (The Hague: TMC Asser Press, 2001), and "Geneva Conventions," at The Free Dictionary website, http://www.thefreedictionary.com.

in particular. But clearly, only a remnant or a feeble custom remained of either after the collapse of the European system at the end of the 19th century. Little was left at the end of World War I, and nothing but a sham at the end of World War II. Schmitt's thesis regarding the situation of international law and the laws of war is undeniable. Moreover, his thesis regarding the transformation of war is also undeniable. Here Schmitt: "Just war, i.e., the de-legitimation of the opponent and the self-empowerment of the just side; that means: transformation of state war (i.e., of war in international law) into a war that is at once colonial war and civil war; that is logical and irresistible: war becomes global civil war and ceases to be interstate war."[27] Moreover: "The terrible transformation of the world, which has been accomplished by a headlong expansion of power, lies in that things beyond the measure of our physically-given sense of perception have been made visible, audible, perceptible; perceptible and thus capable of possession. The new concept of ownership or, much more: the domination of functions; *cuius regio, eius economia*, now *cuius economia, eius regio*. That is the new *Nomos* of the earth; no more *Nomos*."[28] That was especially clear to Schmitt with the onset of the Cold War: "Berlin lies in the flight path between New York and Moscow; in this flight path West and East meet. But these lines have no orientation and no order, and to demonstrate this is precisely the meaning of my *Nomos* of the Earth."[29]

No new *nomos* of the earth can obtain without common agreement on the laws of war. Carl Schmitt understood this very well, which is why the concluding section of *The* Nomos *of the Earth* contains a chapter on "Transformation of the Meaning of War" and ends with a chapter on "War with Modern Means of Destruction." It is thus quite consistent that Schmitt concludes

27. Schmitt, *Glossarium*, op. cit., dated October 8, 1947, p. 29.

28. Schmitt, *Glossarium*, op. cit., dated July 16, 1948.

29. Schmitt, *Glossarium*, op. cit., dated August 20, 1948.

Theory of the Partisan with the statement: "The theory of the partisan flows into the question of the concept of the political, into the question of the real enemy and of a new *nomos* of the earth." As Schmitt notes, his theory of the partisan is not simply another corollary on the concept of the political,[30] but rather consideration of a specific and significant phenomenon that in the second half of the 20th century ushered in a new "*theory* of war and enmity." By the same token, the theory of the terrorist at the end of the 20th and beginning of the 21st century has ushered in a new theory of war and enmity. Schmitt's theory of the partisan immediately raises the question of the distinction between the partisan and the terrorist that logically culminates in the theory of the terrorist, which again flows into the question of the concept of the political, into the question of the *absolute* enemy, i.e., the foe, and of a new *nomos* of the earth in the 21st century.

G. L. Ulmen

30. These corollaries appear in Carl Schmitt, *Der Begriff des Politischen: Text von 1932 mit einem Vorwort und drei Corollarien* (Berlin: Duncker & Humblot, 1963), pp. 97–124.

FOREWORD

This discussion, *Theory of the Partisan,* originated in two lectures I delivered in 1962: on March 15, in Pamplona, at the invitation of General de Navarra College; and on March 17, at the University of Saragossa, at meetings organized by the General Palafox School, at the invitation of its director, Professor Luis García Arias. The lectures were published by the Palafox Professorship at the end of 1962. The subtitle, *Intermediate Commentary on the Concept of the Political,* has reference to the concrete moment of publication, when the publisher [of *The Concept of the Political*] was making my 1932 work available again. In the last decades, a number of corollaries on the subject have appeared. This discussion is not such a corollary, but rather an independent, if sketchy work, whose theme unavoidably flows into the problem of the distinction between friend and enemy. Thus, I would like to offer this elaboration on my 1962 lectures in the modest form of an intermediate commentary, and in this way to make it accessible to all those who have followed attentively the difficult discussion of the concept of the political.

February 1963 Carl Schmitt

INTRODUCTION

View of the Initial Situation 1808–13

The initial situation for our consideration of the problem of the partisan is the guerrilla war that the Spanish people waged against the army of a foreign conqueror from 1808 until 1813. In this war, a people—a pre-bourgeois, pre-industrial, pre-conventional nation—for the first time confronted a modern, well-organized, regular army that had evolved from the experiences of the French Revolution. Thereby, new horizons of war opened, new concepts of war developed, and a new theory of war and politics emerged.

The partisan fights irregularly. But the difference between regular and irregular battle depends on the precision of the regular, and finds its concrete antithesis and thereby its concept only in modern forms of organization, which originated in the wars of the French Revolution. During all ages of humanity and its many wars and battles, there have been rules of war and battle, and thus also transgression and disregard of those rules. In particular, during times of dissolution, e.g., the Thirty Years War on German soil (1618–48) and all civil and colonial wars in world history, there have been recurring phenomena that can be called partisan. Yet, for a complete theory of the partisan, it is important to recognize that the power and significance of his irregularity has been dependent on the power and significance of the regularity that he challenges. Precisely this regularity of the state, as well as of the army, obtained a new, exact resolution in the French state, as well as in the French army through Napoleon. The countless

Indian wars of the white conquerors against the American redskins from the 17th to the 19th century, as well as the methods of the riflemen in the American War of Independence against the regular English army (1774–83), and the civil war in the Vendée between Chouans and Jacobins (1793–96) all still belong to the pre-Napoleonic stage. The new art of war of Napoleon's regular army originated in the new, revolutionary form of battle. A Prussian officer from that time saw Napoleon's whole campaign against Prussia in 1806 as merely "partisan warfare on a grand scale."[1]

The partisan of the Spanish guerrilla war of 1808 was the first who dared to fight irregularly against the first modern, regular army. Napoleon had defeated the regular Spanish army in the fall of 1808; the actual Spanish guerrilla war began only after this defeat of the regular army. There is not yet a complete, documented history of the Spanish partisan war.[2] This history,

1. Eberhard Kessel, "Die Wandlung der Kriegskunst im Zeitalter der französischen Revolution," in *Historische Zeitschrift*, Vol. 148 (1933), pp. 248f., and Vol. 191 (1960), pp. 397ff., which is a review of Robert S. Quimby, *The Background of Napoleonic Warfare: The Theory of Military Tactics in Eighteenth Century France* (New York: Columbia University Press, 1957); Werner Hahlweg, *Preussische Reformzeit und revolutionärer Krieg*, Supp. 18 of the *Wehrwissenschaftlichen Rundschau* (Sept. 1962), pp. 49–50: "Napoleon thereby (through the new form of battle of the revolutionary mass national liberation army) had created an almost perfectly completed system: his operations of great war, his great tactics, and his great strategy." The Prussian officer and publicist Julius von Voss has argued that Napoleon's entire 1806 campaign "could be called a partisan mobilization on a grand scale," p. 14.

2. In the publications of the General Palafox School, see *La Guerra Moderna* (1955): Fernando de Salas Lopez, *Guerrillas y quintas columnas* (Vol. II, pp. 181–211); in *La Guerra de la Independencia Española y los Sitios de Zaragoza* (1958): José Maria Jover Zamora, "La Guerra Española de la Independencia en el Marco de las Guerras Europeas de Liberación (1808–1814)," pp. 41–165; Fernando Solano Costa, "La Resistencia Popular en la Guerra de la Independencia: Los Guerrilleros," pp. 387–423; Antonio Cerrano Montalvo, "El Pueblo en la Guerra de la Independencia: La Resistencia en las Ciudades,"

as Fernando Solano Costa has observed,[3] is necessary, but also very difficult, because the entire Spanish guerrilla war consisted of nearly 200 small, regional wars in Asturias, Aragon, Catalonia, Navarra, Castile, etc., under the leadership of numerous fighters, whose names are shrouded in many myths and legends. One of them, Juan Martín Díez, known as Empecinado, became the terror of the French, and made the road from Madrid to Saragossa unsafe.[4] This partisan war was waged on both sides with the

pp. 463–530. The two fundamental essays by Luis Garcia Arias can be found in *La Guerra Moderna y la Organización Internacional* (Madrid: Instituto de Estudios Politicos, 1962), Vol. I ("Sobre la Licitud de la Guerra Moderna") and in *Defensa Nacional* (1960), "El Nuevo Concepto de Defensa Nacional." Costa concludes his essay with the remark that there is still no documented history of the Spanish national movement against Napoleon. Yet, we must mention in particular his essay (as well as the essay by Zamora) as an outstanding summary, and call special attention to it as an important source of our information. Spanish historical works deal with the guerrilla wars in various ways, and yet not so as to provide contemporary readers with a sufficiently comprehensive picture (Conde de Toreno, Modesto Lafuente, Vol. 5, Rodriguez de Solis, José M. Garcia Rodriguez); the most comprehensive is José Gomez de Arteche in Vols. 4, 5, 7, 9, 11, and 14 of his history of the War of Independence. It is beyond the scope of this investigation to discuss French, English, and German works. See the excellent overview by Fernando Solano Costa, "El Guerrillo y su Trascendencia," in the publications of the Congreso Historico Internacional de la Guerra y su Epoca of the Institucion Fernando el Catolico, Saragossa (March/April 1959); see also in the same report, Santiago Amado Loriga, "Aspectos Militares de la Guerra de la Independencia," and Juan Mercader Riba, "La Organizacion administrativa Francesca en España."

3. In the notes to Costa, "La Resistencia Popular en la Guerra de la Independencia: Los Guerrilleros," ibid.

4. See Costa, "La Resistencia Popular en la Guerra de la Independencia: Los Guerrilleros," op. cit., pp. 387, 402, and 405; Gregorio Marañon has published a Spanish translation of the section on Empecinado in Frederick Hardman, *Peninsular Scenes and Sketches* (Edinburgh and London: W. Blackwood & Sons, 1846). [See *El Empecinado visto por un inglés* (Buenos Aires: Espasa-Caalpe, 1946), with a prologue by Marañon.] José de Arteche has published a lecture on Empecinado in Vol. 14 as an appendix. Along with Empecinado, Priest Merino, to whom the last account in Marañon's translation of Hardman's

utmost cruelty, and it is no wonder that more historical materials, books, and memoirs have been published by the educated *afrancesados* [friends of the French] than by the guerrillas. But, as always, myths and legends, on the one hand, and documented history, on the other, may be complementary; in any case, the lines of our initial situation are clear. According to Clausewitz, half of the entire French army often was posted in Spain, and half of this number, namely 250–260,000 men, were engaged with the guerrillas, whose number has been estimated by Gomez de Arteche at 50,000, and by others as being much lower.

Primarily important in the situation of the Spanish partisan in 1808 was the fact that he risked battle on his own home soil, while the king and his family were not yet able to tell exactly who was the real enemy. In this respect, the legitimate authority in Spain did not act differently from the one in Germany. Another feature of the Spanish situation was that the educated strata of the nobility, the high clergy, and the bourgeoisie were mostly *afrancesados,* i.e., they sympathized with the foreign conqueror. Also in this respect, there are parallels with Germany, where the great German poet Goethe wrote hymns to Napoleon's glory, and the German educated elite never was certain about where it belonged. In Spain, the guerrilla dared the hopeless fight; [he was] a poor devil, the first typical case of the irregular cannon fodder of global political conflicts. All of this is part of an overture to a theory of the partisan.

At that time, a spark jumped from Spain to the North. It did not ignite the same fire that gave the Spanish guerrilla war its world-historical significance. But it produced an effect whose

chapter on Empecinado is dedicated, should be mentioned. In 1823, when the French invaded Spain at the order of the Holy Alliance, Empecinado and Priest Merino stood on opposing fronts: Empecinado, on the side of the constitutionalists; Priest Merino, on the side of the absolutist restoration and the French.

continuation today, in the second half of the 20th century, changes the face of the earth and her humanity. It brought about a *theory* of war and enmity that logically culminates in the theory of the partisan.

At first, in 1809, during the short war waged by the Austro-Hungarian Empire against Napoleon, a systematic attempt was made to emulate the Spanish model. With the help of famous publicists, among them Friedrich Genz and Friedrich Schlegel, the Austrian government in Vienna staged a national propaganda campaign against Napoleon. Spanish writings were disseminated in German.[5] Heinrich von Kleist quickly joined in this anti-French propaganda, and continued it in Berlin after this Austrian war of 1809. In these years, until his death in November 1811, he became the real poet of national resistance against the foreign conqueror. His play *Die Hermannsschlacht* [The Battle of Hermann][6] is the greatest partisan poem of all times. He also wrote a poem titled "An Palafox" [To Palafox], and put the defender of Saragossa in the same company as Leonidas, Arminius, and William Tell.[7] The well-known fact that the reformers in the

5. Peter Rassow discusses the pamphlet by the Spanish Minister Ceballo, Ernst Moritz Arndt, and Kleist's "Katechismus der Deutschen," in "Die Wirkung der Erhebung Spaniens auf die Erhebung gegen Napoleon I," in *Historische Zeitschrift* 167 (1943), pp. 310–35; cf. Hahlweg, *Preussische Reformzeit und revolutionärer Krieg*," op. cit., p. 9, n. 9–13 (on the insurrections in Germany 1807–1813). Colonel von Schepeler, who became known later as a historiographer of the Spanish War of Independence, collaborated on an Austrian plan for armed insurrection: Hans Jureschke, "El Colonel von Schepeler, Caracter y Calor informativo de su obra historiografica sobre el reinado de Fernando VII," in *Revista de Estudios Politicos,* No. 126 (special issue on the constitution of Cadiz 1812), p. 230.

6. [Tr. Heinrich von Kleist, *Die Hermannsschlacht: Ein Drama in fünf Aufzügen*, intro. and annot. Adolf Lichtenfeld (Vienna: Karl Graeser Verlag, 1885). There appears to be no English translation of this work.]

7. Rudolf Borchardt has included Kleist's poem, "An Palafox," in his collection, *Ewiger Vorrat deutscher Poesie* (1926). However, the defender of Saragossa,

Prussian General Staff, most notably August von Gneisenau and Gerhard von Scharnhorst, were deeply impressed by the Spanish example will be discussed below. The origins of the book *On War*,[8] through which the name Carl von Clausewitz acquired an almost mythical resonance, also can be found in the world of ideas of these Prussian staff officers from 1808–1813. His formula of "war as the continuation of politics" already contains in a nutshell a theory of the partisan, whose logic has been pursued to its end by Lenin and Mao Tse-tung.

The only real guerrilla war of national liberation that should be mentioned in the context of our partisan problem was waged in Tyrol, where Andreas Hofer, Joseph Speckbacher, and the Capuchin friar Joachim Happinger were active. The Tyroleans became "a mighty torch," as Clausewitz put it.[9] However, the 1809 episode ended quickly. There was no partisan warfare against the French in the rest of Germany. The strong national impulse, manifested in individual revolts and raiding parties, ended very quickly, and had no impact on the course of regular war. The conflicts of spring and summer 1813 took place on the battlefield, and the decisive blow was struck in a pitched battle near Leipzig in October 1813.

General Palafox, was no partisan, but rather a regular professional officer, and the defense of the town by the entire population, men and women, was still no partisan warfare, but rather regular resistance against a regular siege, as Hans Schomerus emphasizes in [a series titled] "Partisanen" in *Christ und Welt*, which began in Vol. II, No. 27 (July 7, 1949) and concluded in No. 38 (September 22, 1949).

8. [Tr. Carl von Clausewitz, *On War*, ed. and tr. Michael Howard and Peter Paret (Princeton: Princeton University Press, 1989). The most recent German edition, which will be mentioned by Schmitt later, is Carl von Clausewitz, *Vom Kriege*, 19th ed., with further critical comments by Werner Hahlweg (Bonn: Ferd. Dümmlers Verlag, 1980).]

9. Carl von Clausewitz, *Politische Schriften und Briefe*, ed. Dr. Hans Rothfels (Munich: Drei Masken Verlag, 1922), p. 217.

In the course of a general restoration, the Congress of Vienna (1814–15) also reestablished the concepts of European laws of war.[10] This was one of the most remarkable restorations in world history. It was so enormously successful that these laws of war regarding bracketed [i.e., limited] continental land war still dominated the military conduct of European land war during World War I (1914–18). Until today, these laws of war are called "classical," and they deserve the name, because they recognize clear distinctions, above all between war and peace, combatants and non-combatants, enemy and criminal. War was waged between states, between regular state armies, and between sovereign bearers of a *jus belli* [right to war], who also in war respected each other as enemies, and did not discriminate against each other as criminals, so that a peace treaty was possible and even constituted the normal, self-evident end of war. Given such a classical regularity—as long as it actually held sway—the partisan could be only a marginal phenomenon, which in fact he was even during World War I.

10. A number of restorations of the Congress of Vienna still are recognized, e.g., the dynastic principle of legitimacy and legitimate monarchy; then, too, the high aristocracy in Germany, the Pontifical State in Italy, and (via the papacy) the Jesuit order. Less known is the great work of the restoration of the *jus publicum Europaeum* [public law of Europe] and its bracketing of land war between European sovereign states, which, at least in textbooks of international law, has continued until today as a "classical" facade. In my book, *The* Nomos *of the Earth in the International Law of the* Jus Publicum Europaeum, tr. G. L. Ulmen (New York: Telos Press, 2003), the disruption caused by wars during the French Revolution and the Napoleonic age is not dealt with as elaborately as it should be, as Hans Rehberger rightly has criticized in his review in *Friedenswarte*, Vol. 50 (1951), pp. 305–14. As an at least partial supplement, I now can refer to Roman Schnur's research on France's ideas and practice of international law from 1789 to 1815, of which only one essay, "Land und Meer," has appeared in *Zeitschrift für Politik* (1961), pp. 11ff. The framework of the restoration of the bracketing of European war also includes the continued neutrality of Switzerland and its *situation unique*. See *The* Nomos *of the Earth*, ibid., pp. 190 and 249ff.

The Horizon of Our Investigation

If I have occasion to speak about *modern* theories of the partisan, then I must emphasize that there actually are no old theories of the partisan as distinguished from *modern* ones. In the classical laws of war of European international law, there was no place for the partisan in the modern sense. He was either, as in the cabinet wars[11] of the 18th century, a *lighter*, especially more mobile, but nevertheless regular type of troop or, being a particularly gruesome criminal, he was simply outside the law: *hors la loi*. As long as the notion of a duel with open weapons and chivalry remained in war, this could not have been otherwise.

With the introduction of compulsory military service, all wars become in principle wars of national liberation, and this soon leads to a situation that is difficult and often even irresolvable for the classical laws of war, such as a more or less improvised *levée en masse* [mass conscription], or the *Freikorps*[12] and the *Francs-tireurs*,[13] which I will discuss later. Fundamentally, in any case, war remains *bracketed*, and the partisan stands outside of this bracketing. The fact that he stands outside of this bracketing

11. [Tr. Cabinet wars were initiated and conducted by sovereign monarchs and their cabinets, i.e., their closest ministerial advisors.]

12. [Tr. *Freikorps*: "Free-Corps," i.e., voluntary troops, who became particularly significant in the German revolution following the end of World War I.]

13. [Tr. *Francs-tireurs* (literally, "French sharpshooters") refers to the irregular groups that harassed the German army during the latter stages of the Franco-German War (1870–71), after the rapid defeat of the main French army. Originally, the *Francs-tireurs* were derived from rifle clubs or unofficial military societies formed in the east of France during the Luxemburg crisis of 1867. Their members wore no uniforms, were well armed, and elected their own officers. In July 1870, at the outbreak of war, the societies were brought under the control of the minister of war and organized for field service. After November 4, by which time the *levée en masse* was in force, they were placed under orders of generals in the field. The severity of German reprisals against them was testimony to the fear and anxiety inspired by the presence of active bands of free-shooters on the flanks and in the rear of the invaders.]

now becomes a matter of his essence and his existence. The modern partisan expects neither law nor mercy from the enemy. He has moved away from the conventional enmity of controlled and bracketed war, and into the realm of another, real enmity, which intensifies through terror and counter-terror until it ends in extermination.

Given the situation of the partisan, two types of war are especially important and in a certain sense even related to it: civil war and colonial war. Today, this context is very specific. Classical European international law pushed both of these dangerous forms of war and enmity to the margins. In the *jus publicum Europaeum* [public law of Europe], war was waged *between* states, and was war that a regular state army waged against another regular state army. Open civil war was considered to be armed insurrection, which was defeated with the help of a state of siege by police and troops, unless it led to recognition of the rebels as belligerents. Colonial war still remains within the purview of the military science of European nations such as England, France, and Spain. Yet, this still does not challenge regular state war as the classical model.[14]

Here, Russia in particular must be mentioned. During the entire 19th century, the Russian army waged many wars against Asiatic mountain peoples, and never restricted itself exclusively to regular army wars, as did the Prussian-German army. Furthermore, the autochthonous partisan war against Napoleon's army is part of Russian history. In the summer of 1812, Russian partisans under military leadership harassed the French army in its advance on Moscow; in fall and winter of the same year, Russian peasants killed the cold and hungry French army in its retreat. All of this did not last much longer than half a year, but it sufficed to become an historical event of great import, admittedly

14. See the pages listed in the subject index of *The* Nomos *of the Earth*, op. cit., under the keywords: civil war, enemy, *justa causa*, and *justus hostis*.

more through its political myth and its different interpretations than through its paradigmatic impact on the theoretical science of war. Here, we must mention at least two different, even conflicting interpretations of the Russian partisan war of 1812: an anarchistic one, based on Michael Bakunin and Peter Kropotkin, which became famous in Leo Tolstoy's novel *War and Peace*, and the Bolshevik utilization in Stalin's tactics and strategy of revolutionary war.

Tolstoy was not an anarchist like Bakunin or Kropotkin, but his literary influence was much greater. His epic *War and Peace* contains more myth-building power than any political doctrine or any documented history. Tolstoy elevated the Russian partisans of 1812 to bearers of the elemental powers of the Russian earth, which shook off the famous emperor Napoleon and his splendid army like some onerous vermin. For Tolstoy, the uneducated, illiterate peasant not only is stronger, but also more intelligent than all strategists and tacticians; above all, he is more intelligent than the great general Napoleon, who becomes a puppet in the hands of historical events. During World War II, Stalin adopted this myth of the indigenous, national partisan against Germany, and put it very concretely into the service of his communist world politics. This constituted an essentially new stage of partisan warfare, which began with Mao Tse-tung.

For thirty years, partisan warfare had been pursued in many parts of the earth. It began in 1927—before World War II—in China and other Asiatic countries, which later defended themselves against the Japanese invasion from 1932 to 1945. During World War II, Russia, Poland, the Balkans, France, Albania, Greece, and other countries became theaters of this type of warfare. After World War II, partisan warfare continued in Indochina, where the Vietnamese communist leader, Ho Chi-minh, and the victor of Dien Bien Phu, General Vo Nguyen Giap, organized it with peculiar effect against the French colonial army. Moreover, partisan warfare continued in Malaya, the Philippines, Algeria, Cyprus under General George Grivas, and Cuba under Fidel Castro and Che Guevara. At present, in 1962, the Indo-Chinese countries of Laos and Vietnam are theaters of partisan warfare, which daily develops new methods of overpowering and outwitting the enemy. Modern technology provides ever stronger weapons and means of destruction, ever more perfected means of transportation and methods of communication for the partisans, as well as for the regular troops who fight them. In the vicious circle of terror and counter-terror, combat against partisans is often only a mirror image of partisan warfare, and time and again the correctness of the old adage—usually cited as Napoleon's order to General Lefèvre on September 12, 1813—proves to be true: in fighting the partisan anywhere, one must fight like a partisan; *il faut opérer en partisan partout où il y a des partisans.*

Some particular questions regarding norms in jurisprudence and international law will be elaborated below.[15] The fundamentals are self-evident; the application to rapidly developing concrete situations is controversial. In recent years, an impressive document of the will to total resistance, and not only the will, but also detailed orders for its concrete execution, has appeared: The Swiss "Everyman's Guide to Guerrilla Warfare," published by the Swiss Association of Non-Commissioned Officers.[16] In 180 pages, it offers instructions for passive and active resistance against a foreign invasion, with detailed suggestions for sabotage, going underground, concealing weapons, surprise attacks, combating spies, etc. The experiences of the last decades are utilized scrupulously. This everyman's modern war guide begins with

15. See pp. 23ff.

16. Hans von Dach, *Der totale Widerstand: Kleinkriegsanleitung für Jedermann*, 2nd ed. (Biel: Zentralsekretariat des Schweizerischen Unteroffiziersverbandes, 1958).

the remark that "resistance to the end" must adhere to the 1907 Hague Convention Respecting the Laws and Customs of War on Land, and to the 1949 Geneva Conventions. *This goes without saying.* It is also not difficult to calculate how a normal, regular army would react to the practical application of these directions for guerrilla warfare (e.g., silent killing of guards by hitting them with an axe[17]), as long as it did not feel it had been defeated.

Partisan: Word and Concept

The brief listing of a few known names and events, with which we have attempted an initial outline of the horizon of our views, demonstrates the immeasurable wealth of the subject and the problem. It thus is advisable to define some features and criteria more precisely, so that the discussion does not become abstract and boundless. One such feature we mentioned at the beginning of our presentation, when we assumed that the partisan is an *irregular* fighter. The regular fighter is identified by a soldier's uniform, which is more of a professional garb, because it demonstrates the dominance of the public sphere. The weapon is displayed openly and demonstratively with the uniform. The enemy soldier in uniform is the actual target of the modern partisan.

Another feature that comes to the fore today is the intense political engagement that distinguishes the partisan from other fighters. The intense political character of the partisan must be kept in mind, because he must be distinguished from the ordinary thief and violent criminal, whose motives are directed toward private enrichment. This conceptual criterion of the *political* character [of the partisan] has (in exact inversion) the same structure as does the pirate in the law of sea war. The concept of [the pirate] has the *unpolitical* character of his evil deeds, which are focused on private robbery and profit. The pirate has,

17. Ibid., p. 43.

as the jurists say, *animus furandi* [evil intent]. The partisan fights at a political front, and precisely the political character of his acts restores the original meaning of the word *partisan.* The word derives from *party*, and refers to the tie to a fighting, belligerent, or politically active party or group. These ties to a party become especially strong in revolutionary times.

In revolutionary war, belonging to a revolutionary party implies no less than total inclusion, and a revolutionary fighting party can integrate its active combatants totally, whereas that is not possible for other groups and associations, in particular the contemporary state. In the comprehensive discussion concerning the so-called total state, it still has not been realized that today it is not the *state*, but rather the revolutionary *party* that represents the actual and in fact the only totalitarian organization.[18] In purely organizational terms, in terms of the strict functioning of protection and obedience, some revolutionary organizations are superior to some regular troops. If organization as such is made a criterion of regularity, as occurred in the 1949 Geneva Conventions,[19] then there will be confusion in the international law of war.

The word *partisan* means one who follows a party, and what that means concretely is very different at different times. This also is true with respect to the party or the front someone belongs to, and also with respect to following, fellow-traveling, comrades-in-arms, and eventually, to becoming fellow-captives. There are belligerent parties, but also parties in judicial processes, parties of parliamentary democracy, parties of opinion and activism, etc. In romance languages, the word *partisan* can be used as a

18. See gloss 3 to "Weiterentwicklung des totalen Staates in Deutschland" (1933), in Carl Schmitt, *Verfassungsrechtliche Aufsätze aus den Jahren 1924–1954: Materialien zu einer Verfassungslehre* (Berlin: Duncker & Humblot, 1958 [2nd ed., 1973]), p. 366.

19. See pp. 25ff.

noun and as an adjective: in French, one even can speak of the *partisan* of some opinion; in short, a very general, ambiguous designation suddenly becomes a highly political word. Linguistic parallels with a general term like *status*, which suddenly can mean *state*, come to mind. In times of dissolution, as in the 17th century during the Thirty Years War, the irregular soldier fell into the company of street robbers and vagrants; he waged war for personal profit, and became a figure of the picaresque novel, like Estebanillo Gonzales' Spanish Pícaro [rogue], who was associated with the battle of Nördlingen (1635) and told about it in the style of the [good] soldier Schweik. One also can read it in Grimmelshausen's *Simplizius Simplizissimus*, and see it in Jean Callot's prints and etchings. In the 18th century, the "party-follower" belonged to the pandours,[20] the hussars,[21] and other types of light troops, which, as mobile troops, "fight separately" and pursue so-called small wars, by comparison with the slower great wars of regular troops. Here, the distinction between regular and irregular is meant in solely military-technical terms, and in no sense is synonymous with legal and illegal in the juridical sense of international and constitutional law. As regards contemporary partisans, the antithetical pairs of regular-irregular and legal-illegal usually become blurred and interchangeable.

Flexibility, speed, and the ability to switch from attack to retreat, i.e., increased mobility, remains today characteristic of the partisan, and this characteristic is even more intensified through technicization and motorization. Both antithetical pairs

20. [Tr. Pandours were members of a local Croatian military force organized in 1741 by Baron Franz von der Trenk from his tenants to repress the brigands on the Turkish frontier. Later, they were enrolled as a regiment in the Austrian army. Their desperate courage, cruelties, and plundering made them dreaded throughout Germany.]

21. [Tr. Originally, hussars (literally, freebooters) were members of the light cavalry in Hungary and Croatia; then, they became a class of calvary of European armies and usually were distinguished by a brilliant and much-decorated uniform.]

[regular/irregular and legal/illegal] are dissolved only in revolutionary war, when numerous semi- and para-regular groups come into being. But the armed partisan remains dependent on cooperation with a regular organization. Fidel Castro's comrade-in-arms in Cuba, Ernesto Che Guevara, is an especially good example.[22] Consequently, some intermediary stages develop through the cooperation of regular and irregular, also in cases in which a by no means revolutionary government calls for defense of the national soil against a foreign invader. A war of national liberation and a guerrilla war then become one. The designation *partisan* has obtained in official calls to arms since the 16th century.[23] We still will become acquainted with two important examples of a formal regulation of wars of national liberation and of local militias that tried to regulate guerrilla warfare. But the foreign invader also issues regulations for combating enemy partisans.

All such uniform codes face the difficult problem of an international legal regulation, i.e., a regulation of the irregular that is valid for both sides regarding, on the one hand, recognition of the partisan as a combatant and his treatment as a prisoner of war, and, on the other, respect for the rights of the occupying military power. We already have mentioned that in this respect there are some juridical controversies, and we will return to the debate concerning the *Francs-tireurs* of the Franco-German War (1870–71) after we have examined the situation in international law.

22. Ernesto Che Guevara, *On Guerrilla Warfare*, intro. Major Harries-Clichy Peterson (New York: Frederick A. Praeger, 1961), p. 9: "It is obvious that guerrilla warfare is a preliminary step, unable to win a war all by itself."

23. Manuel Fraga Iribane, in his essay "Guerra y Politica en el siglo XX," points out that since 1595 there have been French decrees regarding enemy invasions in which the terms *partisan* and *parti de guerre* are used. Cf. *Las Relaciones Internacionales en la era de la guerra fria* (Madrid: Instituto de Estudios Politicos, 1962), p. 29, n. 62 and 27.

In general, and in view of the rapid changes in the world, the tendency of traditional or "classical" concepts, as one likes to call them today, to be changed or also to be dissolved is all too understandable.[24] This is also the case with the "classical," if one may call it that, concept of the partisan. In a very important book for our subject, the illegal resistance fighter and underground activist are made the true type of partisan.[25] This is a conceptual transformation oriented primarily to certain inner-German situations of the Hitler time, and has this aim in view. Illegality is substituted for irregularity; resistance, for military combat. It appears to me that this implies a wide-ranging re-interpretation of the partisan of wars of national liberation, and a failure to recognize that even the revolutionizing of war has not forsaken the military connection between the regular soldier and the irregular fighter.

In some cases, the re-interpretation leads to a general symbolization and dissolution of the concept. Then, ultimately, any individualist and non-conformist can be called a partisan, without any consideration as to whether he would even think of taking up arms.[26] As a metaphor, this may be permissible, but for me the term *partisan* defines specific historical figures and

24. See my lecture, "El orden del mundo después de la segunda guerra mundial," in *Revista de Estudios Politicos* (1962), No. 122, p. 12, and the keyword "classical" in the subject index of Schmitt, *Verfassungsrechtliche Aufsätze*, op. cit., p. 512.

25. Rolf Schroers, *Der Partisan: Ein Beitrag zur politischen Anthropologie* (Cologne: Kiepenheuer und Witsch, 1961). In the course of our discussion, we will return several times to this especially important book; cf. p. 47, n. 16, where Schroers rightly distinguishes the partisan from the revolutionary agent, the functionary, the spy, and the saboteur, but identifies him with the resistance fighter in general. In contrast, I adhere to the criteria mentioned in the text, and hope thereby to have taken a clearer position, which makes possible a fruitful debate.

26. See Hans Joachim Sell, *Partisan* (Düsseldorf: Eugen Friedrichs Verlag, 1962). This is a superb novel, with psychologically and sociologically

situations.[27] In a general sense, we may say "to be a man is to be a fighter," and the consistent individualist does indeed fight on his own terms and, if he is courageous, at his own risk. He then becomes his own party-follower. Such conceptual dissolutions are noteworthy signs of the times, which deserve a separate examination.[28] Yet, for a theory of the partisan as it is meant here,

interesting depictions of aristocratic and bourgeois figures in the Federal Republic of Germany during the 1950s.

27. Thus, in an essay on Lorenz von Stein in 1940 ["Die Stellung Lorenz von Steins in der Geschichte des 19. Jahrhunderts," in *Schmollers Jahrbuch für Gesetzgebung, Verwaltung, und Volkswirtschaft im Deutschen Reiches*, Vol. 64, No. 6, pp. 641–46] and in a lecture on Donoso Cortés in 1944, I called Bruno Bauer and Max Stirner "partisans of the world spirit." See "Donoso Cortés in gesamteuropäischer Interpretation," in *Donoso Cortés in gesamteuropäischer Interpretation—Vier Aufsätze* (Cologne: Greven Verlag, 1950), p. 100. In an essay on the occasion of the 250th anniversary of Rousseau's death, "Dem wahren Johann Jakob Rousseau—Zum 28. Juni 1962," in *Zürcher Woche*, No. 26 (June 29, 1962), p. 1, in reference to Schroers and Sell, I have used the figure of the partisan to clarify the controversial image of Rousseau. In the meantime, I have become familiar with an essay by Henri Guillemin, "J. J. Rousseau, trouble-fête," which seems to confirm this interpretation. Guillemin is the editor of Rousseau's *Lettres écrites de la Montagne* (Neuchatel: Collection du Sablier, editions Ides et Calendes, 1962), with an important prologue.

28. Whereas Schroers, *Der Partisan: Ein Beitrag zur politischen Anthropologie*, op. cit., sees in the partisan the last resistance against the nihilism of a thoroughly technologized world, the last defender of species and soil, and ultimately, the last man, to Gerhard Nebel the partisan appears to be precisely the opposite, i.e., a figure of modern nihilism, who, like the fate of our century, encompasses all professions and classes: the priest, the peasant, the scholar, and, in this way, also the soldier. Nebel's book, *Unter Partisanen und Kreuzfahrern* (Stuttgart: Ernst Klett Verlag, 1950), is the war journal of a German soldier in 1944/45 in Italy and Germany, and it would be worth the effort to compare his depiction of the partisan in Italy at that time with Schroers' interpretation (cf. p. 243). In particular, Nebel's book contains a remarkable account of the moment in which a large regular army dissolves and, as a mob, either is killed by the population or itself kills and plunders, whereby both sides could be called partisan. If Nebel, beyond his good descriptions, categorizes the poor devils and rogues as "nihilists," this is only a metaphysical spice of the time, and today belongs to that time, just as the picaresque novel of

a number of criteria must be kept in mind, so that the theme does not dissolve into abstract generalities. Such criteria are: irregularity, increased mobility of active combat, and increased intensity of political engagement.

I would like to retain a further, fourth characteristic of the true partisan, which Jover Zamora has called the telluric. Despite all tactical mobility, this characteristic is important for the basically defensive situation of the partisan, who changes his essence once he identifies with the absolute aggressivity of a world-revolutionary or a technicistic ideology. Two especially interesting discussions of the theme, the book by Rolf Schroers[29] and the dissertation by Jürg H. Schmid[30] about the position of the partisan in international law, basically correspond to these criteria. Zamora's stress on the telluric character [of the partisan] seems to me necessary in order to make the defensive, i.e., the limited nature of hostility, spatially evident, and to guard it against the absolute claim of an abstract justice.

For the partisans who fought in Spain (1803–1813), Tyrol, and Russia, this is self-evident. But also the partisan struggles

the 17th century belongs somewhat to scholastic theology. Ernst Jünger, *Der Waldgang* (Frankfurt a/M: Verlag Vittorio Klostermann, 1951), constructs the *Waldgänger* [the walker in the forest], whom he calls partisan several times, as a "figure" in the sense of the "worker" (1932). The individual, surrounded by machines, does not give up in the seemingly desperate situation, but wants to continue with utmost fortitude, and "decides to walk in the forest." "As for place, the forest is everywhere" (p. 11). Gethsemane, for example, the Mount of Olives, which we know from the passion story of Jesus Christ, is "forest" in Jünger's sense (p. 73), but also the Daimonion of Socrates (p. 82). Accordingly, the "teachers of law and constitutional law" are denied their ability to "give the walker in the forest the necessary means of defense. Poets and philosophers already see better what is needed" (p. 126). Only theologians know the true sources of power. "Because all wise men understand the theologians" (p. 95).

29. Schroers, *Der Partisan: Ein Beitrag zur politischen Anthropologie*, op. cit.

30. Cf. Jürg H. Schmid, "Die völkerrechtliche Stellung der Partisanen im Kriege," in *Zürcher Studien zum Internationalem Recht*, No. 23 (1956).

during World War II and the years thereafter in Indochina and other countries, well identified by the names of Mao Tse-tung, Ho Chi-minh, and Fidel Castro, indicate that the tie to the soil, to the autochthonous population, and to the geographical particularity of the land—mountain-ranges, forests, jungles, or deserts—are topical even today. The partisan is and remains distinct, not only from the pirate, but likewise from the corsair,[31] even as land and sea, as different elemental spaces of human labor and military struggle between nations, remain distinct. Land and sea have developed not only different means of pursuing war, and different theaters of war, but also different concepts of war, enemy, and booty.[32] For at least as long as anti-colonial wars are possible on our planet, the partisan will represent a specifically terrestrial type of active fighter.[33] The telluric character of the partisan will

31. [Tr. Corsair was the name given by the Mediterranean peoples to the privateers of the Barbary coast, who plundered the shipping of Christian nations. Strictly speaking, they were not pirates, since they were commissioned by their respective governments. Even so, the word came to be synonymous with "pirate" in English (but not in French).]

32. Carl Schmitt, *Land und Meer—Eine weltgeschichtliches Betrachtung* [1942/1954] (Hohenheim: Edition Maschke, 1981); *The* Nomos *of the Earth*, op. cit., pp. 172ff. and 310; "Die geschichtliche Struktur des heutigen Weltgegensatzes von Ost und West," in *Freundschaftliche Begegnungen: Festschrift für Ernst Jünger zum 60. Geburtstag* (Frankfurt a/M: Vittorio Klostermann, 1955), pp. 135–67. In this latter essay, which simultaneously appeared in *Revista de Estudios Politicos*, No. 81 (Madrid, 1955), I announced that I would like to develop fully and hermeneutically sections 247–48 of Hegel's *Philosophy of Right* as a historico-intellectual nucleus for understanding the contemporary techno-industrial world, just as the Marxist interpretation developed the preceding sections 243–46 for an understanding of bourgeois society.

33. In her review of Schroers' book, *Der Partisan: Ein Beitrag zur politischen Anthropologie*, op. cit., Margret Boveri (in *Merkur*, No. 168 [February 1962]) praises Czeslav Milosz's book, *West- und Östliches Gelände* (Cologne: Kiepenheuer und Witsch, 1961). The author presents a vivid and sympathetic picture of his life in Lithuania, Poland, and Western Europe, especially Paris, and tells of his underground existence in Warsaw during the German occupation, where he disseminated pamphlets against the Germans. He says explicitly that

be elucidated below through a comparison with typical figures in maritime law[34] and a discussion of the spatial aspect.[35]

Yet, even the autochthonous partisan of agrarian background is being drawn into the force-field of irresistible, techno-industrial progress. His mobility is increased by his motorization to such an extent that he is in danger of becoming completely disoriented. In the situations of the Cold War, he becomes a technician of the invisible struggle, a saboteur, and a spy. Already during World War II, there were sabotage troops with partisan training. Such a motorized partisan loses his telluric character and becomes only the transportable and exchangeable tool of a powerful central agency of world politics, which deploys him in overt or covert war, and deactivates him as the situation demands. This possibility is also part of his present-day existence, and should not be neglected in a theory of the partisan.

With these four criteria—irregularity, increased mobility, intensity of political engagement, and telluric character—and with reference to the possible impact of further technicization, industrialization, and deagrarianization, we have outlined conceptually the horizon of our investigation. It ranges from the guerrilla of Napoleonic times to the well-equipped partisan of the present day, from Empecinado via Mao Tse-tung and Ho Chi-minh to Fidel Castro. This is a large field, from which historiography and the science of war have elaborated an enormous amount of material that expands daily. We will utilize it as far as it is accessible to us, and try to gain some insights for a theory of the partisan.

he was not a partisan and that he did not want to be one (p. 276). Yet, his love for his Lithuanian homeland and its forests does support us in adhering to the telluric character of the partisan.

34. See p. 29.

35. See p. 68.

View of the Situation in International Law

The partisan fights irregularly. But some categories of irregular fighters are equated with regular troops, and enjoy the same rights and privileges. That means: their acts of combat are not illegal, and if they are captured by their enemy, they have a right to special treatment as prisoners of war and as wounded. The legal situation was summarized in the 1907 Hague Convention, which today is accepted generally as valid. After World War II, the development was continued in the 1949 Geneva Conventions, two of which regulate the fate of wounded and sick in land war and sea war, the third regulates treatment of prisoners of war, and the fourth regulates protection of civilians during wartime. Numerous states of both the Western world and the Eastern bloc have ratified them, and their formulations also have been adopted in the new American military handbook of land war law (July 18, 1956).

The 1907 Hague Convention, under certain conditions, equated militia, volunteer corps, and comrades-in-arms of spontaneous insurrections with regular troops. Later, in our discussion of Prussian incompatibility with partisan warfare, we will mention some difficulties and confusions caused by this regulation. The development that led to the 1949 Geneva Conventions is characteristic of the further loosening of the traditional European international law based on the state. Ever more categories of belligerents now are considered to be combatants. Civilians of territories militarily occupied by the enemy (i.e., those in the actual combat zone of partisans fighting behind enemy lines) now also enjoy greater legal protection than those accorded by the 1907 Hague Convention. Many comrades-in-arms, who were considered to be partisans, now are equated with regular combatants, and have rights and privileges. In fact, they no longer can be called partisans. Yet, the concepts still are unclear and vague.

The formulations of the 1949 Geneva Conventions have European experiences in view, but not Mao Tse-tung's partisan warfare and the later development of modern partisan warfare. In the years immediately after 1945, it was not yet realized what an expert like Hermann Foertsch formulated this way: belligerent actions after 1945 had assumed a partisan character, because those who had nuclear bombs shunned using them for humanitarian reasons, and those who did not have them could count on these reservations—an unexpected effect of both the atomic bomb and humanitarian concerns. The norms of the 1949 Geneva Conventions, which are important for the problem of the partisan, are abstracted from particular situations. They are a precise reference, "*une reference précise*,"[36] to the resistance movements of World War II (1939–45).

A fundamental change in the 1907 Hague Convention went unnoticed. The four classical conditions for an equation with regular troops—responsible officers, firm and visible symbols, open display of weapons, observance of rules and application of laws of war—were upheld rigidly. The convention for protection of civilian populations certainly should be valid not only for wars between states, but for all international armed conflicts, and also for civil wars, uprisings, etc. Yet, only the legal foundation for humanitarian interventions of the International Committee of the Red Cross (and other non-partisan organizations) should be created. *Inter arma caritas* [between the arms of charity]. In Art. 3, Sec. 4 of the convention, it is stated expressly that the legal position, *le statut juridique*, of the conflicting parties will not be affected.[37] In wars between states, the occupying power of the militarily occupied area still has the right to direct the local police

36. As stated in the seminal commentary of the International Red Cross led by Jean S. Pictet, Vol. III (1958), p. 65.

37. Ibid., Vol. III, pp. 39–40.

of this area to maintain order and to suppress irregular hostile actions. By the same token, it also has the right to prosecute partisans "without regard to which ideas inspired them."[38]

Accordingly, the distinction of partisans—in the sense of irregular fighters, *not* equated with regular troops—even today remains fundamentally true. The partisan in this sense does *not* have the rights and privileges of combatants; he is a criminal according to ordinary law, and should be made harmless with summary punishment and repressive measures. That was fundamentally recognized also in the war crimes trials after World War II, namely in the Nuremburg judgments against German generals (Alfred Jodl, Wilhelm von Leeb, and Wilhelm List), whereby it was self-understood that, beyond the necessary struggle against partisans, this included all the atrocities, terroristic measures, collective punishments, or even participation in genocide and war crimes.

The 1949 Geneva Conventions widened the circle of persons equated with regular fighters, above all in that members of an "organized resistance movement" were equated with members of militias and voluntary corps, and in this way were granted the rights and privileges of regular combatants. There was no express stipulation regarding a military organization (Art. 13, the convention on the wounded; Art. 4, the convention on prisoners-of-war). The convention on the protection of civilian populations mentioned "international conflicts," which would be settled by force of arms, the same as wars between states in classical European international law, confirming thereby the core of a typical legal institution—*occupatio bellica* [military occupation]—for the traditional laws of war. To such expansions and relaxations, which here can be indicated only by way of example, are added the great transformations and changes that result from

38. Ibid., Vol. IV (1958), p. 330.

development of modern weapons technology, which have a still more intense effect on partisan warfare. For example, what is the meaning of the rule that weapons must be "carried openly," when a resistance fighter in the Swiss "Everyman's Guide to Guerrilla Warfare" says: "Move only at night and rest in the woods during the day"?[39] Or what is the meaning of the requirement of a clearly visible badge of rank during combat at night or in a contest between long-range weapons of modern technical warfare? Many such questions come to the fore, if the discussion is considered from the viewpoint of the problem of the partisan,[40] and if aspects of spatial changes and techno-industrial development are considered.

Protection of civilian populations in militarily occupied areas is protection in many respects. The occupying power has an interest in seeing that peace and order obtain in the militarily occupied area. The population of the occupied area is obligated, certainly not to fidelity, but rather to obedience with respect to the regulations of the occupying power that are in accord with the laws of war. Even civil servants (including the police) should continue to work correctly, and correspondingly should be handled correctly by the occupying power. The whole matter is an extremely difficult balanced compromise between the interests of the occupying power and those of the enemy. The partisan disturbs this type of order in the occupied area in a dangerous way—not only because his proper field of battle is the area behind the enemy's front lines, where he upsets transport and supply, but also if he is more or less supported and hidden by the indigenous population. "The population is your best friend," according to the "Everyman's Guide to Guerrilla Warfare."[41] Protection of such an indigenous population then is potentially also

39. Dach, *Der totale Widerstand*, op. cit., p. 33.

40. Ibid., pp. 71 and 79.

41. Ibid., p. 28.

protection of the partisan. This explains why, in the history of development of the laws of land war, in the deliberations of the 1907 Hague Convention and its further development, a typical oppositional grouping always enters the picture: the great military powers, namely, the potential occupying powers, demanded strict rules of order in the militarily occupied area, whereas the smaller states, i.e., those that feared becoming militarily occupied—Belgium, Switzerland, and Luxembourg—sought at least further protection of resistance fighters and the civilian population. Also in this respect, developments since World War II have led to new perceptions,[42] and destruction of social structures has raised the question of whether there also can be cases in which the population needs protection from the partisans.

The 1949 Geneva Conventions have introduced changes in the precisely-regulated classical legal institutions of *occupatio bellica* of the 1907 Hague Convention, changes whose repercussions remain obscure in many respects. Resistance fighters, who earlier were treated as partisans, are equated with regular troops if they are *organized*. As opposed to the interests of the occupying power, the interests of the occupied territory's indigenous population are stressed so strongly that it becomes possible (at least in theory) that any resistance to an occupying power, also that of partisans, as long as it springs from respectable motives, is considered to be *not illegal*. Nevertheless, the occupying power is justified in using repressive measures. In this situation, a partisan does not become actually legal, but also not actually illegal. He proceeds at his own risk, and, in this sense, is treated as being *risky*.

If one uses words such as *risk* and *risky* in a general, i.e., imprecise sense, then one must conclude that, in an area militarily occupied by enemies and riddled with partisans, in no way is it only the partisan who lives at risk. In this general sense of

42. Ibid., p. 51.

insecurity and danger, the area's entire population is at great risk. Officials who, in accord with the 1907 Hague Convention, want to continue to do their duty correctly, still face an additional risk for acting and failing to act. In particular, police officials are put in a position of dangerous and contradictory expectations: the enemy occupying power demands obedience in accord with the maintenance of security and order, which is precisely what is disturbed by the partisans; the nation-state proper demands loyalty and, after the war, will hold them responsible; the population to whom they belong expects loyalty and solidarity, which, with respect to the activity of police officials, can lead to completely contradictory practical consequences, if police officials are not of a mind to become partisans; and finally, partisans, as well as their opponents, quickly become caught in the vicious circle of reprisals and anti-reprisals. Generally speaking, risky acts and failures to act are not a specific characteristic of partisans.

Thus, the word *risky* also has a precise meaning, namely, that risky actions are treated at their own risk, and the worst consequences of their success or failure are taken for granted, so that there can be no question of injustice when the severest consequences ensue. By the same token, to the extent that risky actions are not treated as unjust acts, there is the possibility that risk can be balanced, that it can culminate in a security treaty. The juridical home of the concept of *risk*—its juridical *topos* [orientation]—remains the right of insurance. Men live with diverse dangers and insecurities, and a danger or insecurity with juridical consequences presents risk, meaning that it makes it and those affected by it *insurable*. With partisans, that might depend on the irregularity and illegality of their actions, even when one might be disposed to protect them technically by classifying them as highly dangerous and all too great a risk.

For situations of war and the confirmation of enmity, consideration of the concept of risk is essential. In Germany, the word *risk* entered the theory of war in international law through a book by Joseph L. Kunz.[43] Yet, risk in this book is not related to land war and certainly not to the partisan. It also does not belong there. Leaving aside the right of insurance as the juridical home of the concept of risk and the imprecise uses of the word, for example, the comparison with escaping prisoners, who risk being shot, Kunz's specific and fruitful use of the concept of *risk* pertains to the laws of sea war, and he has typical figures and situations in mind. Sea war is largely trade war; it has, as opposed to land war, its own space and its own concepts of enmity and booty. Even amelioration of the wounded has led to two conventions—one for land and another for sea—in the 1949 Geneva regulations.

In sea war, "risky" used in such a specific sense refers to two participants: the neutral blockade runner and the neutral contraband leader. With reference to both, the word is precise and meaningful. Both types of belligerents get involved in a "very profitable, yet very risky commercial venture":[44] they risk ship and freight in case they are caught. Thus, they have no enemy, even if they are treated as an enemy in the sense of sea war law. Their social ideal is good business. Their field of operations is the free sea. They do not think about protecting house, herd, or home against a foreign invader, which is the archetype of the autochthonous partisan. They also conclude insurance treaties to compensate for their risk, whereby the insurance costs are correspondingly high, as are the changing risk factors—for example, being sunk by a submarine: very risky, yet highly insured.

One should not take such an apt word as *risky* out of the conceptual field of sea war law and let it become lost in an obscure general concept. For us, since we adhere to the telluric character of the partisan, this is especially important. If earlier, I

43. Joseph L. Kunz, *Kriegsrecht und Neutralitätsrecht* (Vienna: J. Springer, 1935), pp. 146 and 274.

44. Ibid. p. 227.

once characterized freebooters and seafarers as "partisans of the sea,"[45] today I would like to correct this imprecise terminology. The partisan has an enemy and risks something completely different from blockade runners and contraband leaders. He risks not only his life, as does every regular combatant; he knows and accepts that he is an enemy outside of right, law, and honor.

That is also true of the revolutionary fighter, who declares his enemy to be a criminal, and that all concepts of right, law, and honor are ideological swindle. From World War II and the postwar period until today, despite all the characteristic links and mixtures of both types of partisans—defensive autochthonous defenders of the homeland and globally aggressive revolutionary activists—the antithesis remains. It is based, as we will see, on fundamentally different concepts of war and enmity, which are realized in different types of partisans. Where war on both sides is pursued as a non-discriminatory fight between states, the partisan is a marginal figure, who does not rupture the structure of war and does not change the total structure of the political process. However, if the war as a whole is fought with criminalizations of opponents (for example, civil war between class enemies), the ultimate goal is destruction of the enemy state's government; then, the revolutionary disruption of criminalization of the enemy follows in such a way that the partisan becomes the true hero of war. He enforces the death penalty against the criminal, and, if the tables are turned, risks being treated as a criminal or a parasite. That is the logic of a war of *justa causa* [just cause] without recognition of a *justus hostis* [just enemy]. Thereby, the revolutionary partisan becomes the true central figure of war.

However, the problem of the partisan becomes the best touchstone. The various types of partisan warfare still are so confused and fused in the praxis of contemporary war; in their

45. Schmitt, *The* Nomos *of the Earth*, op. cit., p. 174.

fundamental presuppositions, they remain so different that they test the criterion of the friend-enemy grouping. Earlier, we mentioned the typical oppositional grouping of the 1907 Hague Convention: the great military powers *vis-à-vis* the small neutral countries. In the deliberations of the 1949 Geneva Conventions, a compromise formula was achieved with great effort, whereby organized resistance fighters were equated with volunteers. Here also, the typical oppositional grouping reappeared when, in the light of the experiences of World War II, the endeavor was made to embrace it in the norms of international law. Here also, the great military powers, the potential occupiers, were pitted against the small states that feared occupation; this time, however, with an equally striking and symptomatic modification: the world's greatest land power, the Soviet Union, which was the strongest potential occupier, stood now on the side of the small states.

In Schmid's substantive and well-documented work,[46] he attempts to follow "guerrilla warfare through civilians" (concretely focused on Stalin's partisans[47]) and to put them "under the shield of law." This is how Schmid sees "the quintessence of the partisan problem" and the creative legal achievement of the 1949 Geneva Conventions. Schmid would like to see "the renewal of remnants of occupation law" still emanating from the power of occupation, in particular the "much praised duty of obedience." To this end, he relies on the theory of the legal, though risky conduct of war, which he accentuates in a risky, but not illegal conduct of war. Thus, he minimizes the risk of partisans, which he writes off as the costs of the occupying power, given its many rights and privileges. How he intends to oppose the logic of terror and counter-terror is unclear to me; it seems that he simply criminalizes the enemy of the partisans. The whole scenario is a highly interesting hybrid of two different *statuts juridiques* [legal

46. Schmid, "Die völkerrechtliche Stellung der Partisanen im Kriege," op. cit.
47. Ibid., pp. 97 and 157.

statuses], namely, combatants and civilians, with two different types of modern warfare, namely, hot war and cold war, between the population and the occupying power, which Schmid's partisan (following Mao) joins *à deux mains* [with both hands]. What is astounding is only—and this is a true conceptual breakdown—that this de-illegalization of Stalin's partisans at the expense of classical international law is simultaneously linked with the Rousseau-Portalis Doctrine of a pure war between states, concerning which Schmid claims that it would have forbidden civilians from engaging in hostile acts, but only "initially." In this way, the partisan becomes insurable.

The 1949 Geneva Conventions are the work of a human disposition and a humanitarian development that are admirable. Given that they give the enemy not only humanity, but even justice in the sense of recognition, they remain based on the foundation of classical international law and its tradition, without which such a work of humanity would be improbable. Its fundament remains the conduct of war based on the state and a bracketing of war, with its clear distinctions between war and peace, military and civilian, enemy and criminal, state war and civil war. However, to the extent that these essential distinctions are loosened or even challenged, the door is opened for a type of war that consciously destroys these clear separations. Then, many discretely stylized compromise norms appear only as the narrow bridge over an abyss, which conceals a successive transformation of concepts of war, enemy, and partisan.

Development of the Theory

Prussian Incompatibility with the Partisan

In Prussia, the leading military power of Germany, the uprising against Napoleon in early 1813 was fueled by a strong national feeling. The great moment passed quickly; however, it remains so essential in the history of partisan warfare that later we must focus on it. First, however, we must take note of an undisputed historical fact, namely, that the Prussian army and the German army led by Prussia from 1813 through the early part of World War II furnished the classical example of a military organization that had repressed radically the idea of the partisan. The thirty years of German colonial domination in Africa (1885–1915) were not important enough militarily to cause the extraordinary theoreticians of the Prussian General Staff to take the problem seriously. The Austro-Hungarian army had to deal with partisan warfare in the Balkans, and had a regulation for guerrilla warfare. By contrast, the Prussian-German army that marched into Russia on June 22, 1941, did not conceive of partisan warfare. Its campaign against Stalin began with the maxim: troops will fight the enemy; marauders will be handled by the police. The first special directives regarding fighting partisans came only in October 1941; in May 1944, scarcely one year before the end of the four-year war, the first complete regulation of the Supreme Command of the armed forces [regarding partisan warfare] was instituted.[48]

48. Schomerus, "Partisanen," in *Christ und Welt*, op. cit., in particular the section titled "Der Wall der Tradition." The following articles by Schomerus in the same journal remain of great significance for the partisan problem.

In the 19th century, the Prussian-German army became the most famous and exemplary military organization in the European world. But it owed this reputation exclusively to military victories over other regular European armies, in particular those of France and Austria. Only during the Franco-German War (1870/71) in France did it encounter irregular warfare, in the form of the *Francs-tireurs*, who in Germany were called snipers and were handled relentlessly according to the laws of war, just as had been done by every regular army. The more severely a regular army is disciplined, the more correctly it distinguishes between military and civil, and only the enemy in uniform is considered to be an enemy. A regular army becomes more sensitive and more nervous when it encounters a non-uniformed, civilian population on the other side of the struggle. Then, the military responds with harsh reprisals, on-site inspections, hostage taking, destruction of villages, and considers this to be the correct self-defense against treachery and perfidy. The more the regular, uniformed opponent is respected as an enemy, and, also in the most bloody struggles, is not considered to be a criminal, the more ruthlessly the irregular fighter is treated as a criminal. That follows from the logic of classical European laws of war, which distinguish between military and civil, combatants and non-combatants, and which summon the rare moral courage not to declare the enemy to be a criminal.

The German soldier met the *Francs-tireurs* in France, in the autumn of 1870, and on September 2 achieved the great victory over Napoleon III's regular army near Sedan. Had the war been fought according to the rules of classical, regular army warfare, one could expect that after such a victory the war would have ended and peace would have been declared. But instead, the vanquished French imperial government was dismissed. The new republican government under Leon Gambetta proclaimed national resistance against the foreign invader: "all-out war." In ever greater haste, it continually conscripted new armies and threw new masses of badly-trained soldiers onto the battlefield. In November 1870, it even had a military success in the Loire Valley. The situation of the German army was threatened, and German foreign policy was endangered, because a long war had not been foreseen.

The French population was aroused with patriotic fervor, and participated in the struggle against the Germans in various forms. The Germans arrested dignitaries and so-called notables as hostages, shot *Francs-tireurs* caught with guns in their hands, and pressured the population with all kinds of reprisals. That was the situation at the outset of more than half a century of struggle between international law jurists and public propaganda on both sides—for and against the *Francs-tireurs*. The controversies flared again in World War I as a Belgian-German *Francs-tireurs* struggle. Whole libraries have been written about the problem, and even in recent years (1958–60) a panel of respected German and Belgian historians sought to clarify and resolve at least one controversial point in this complex problem of the Belgian *Francs-tireurs* struggle of 1914.[49]

All this is illuminating for the problem of the partisan, because it demonstrates that a normative regulation—if it is conceived to be a factual state of affairs, rather than just a collection of value judgments and general clauses—is juridically impossible. The traditional European bracketing of wars between states emerged after the 18th century from specific concepts of bracketed war and just enemy derived from the age of monarchy. These concepts were interrupted by the French Revolution, but the Congress of Vienna reaffirmed them and they became

49. E. Kessel, in *Historische Zeitschrift*, Vol. 191 (October 1960), pp. 385–93; Franz Petri and Peter Schoeller, "Zur Bereinigung des Franktireurproblems vom August 1914," in *Vierteljahreshefte für Zeitgeschichte*, Vol. 9 (1961), pp. 234–48.

thereby much stronger. But they became legalized between states only when the belligerent states—both internally and externally—adhered to them in equal measure, i.e., when their domestic and foreign policy concepts of regularity and irregularity, legality and illegality, became substantively congruent or at least more or less homogeneous in structure. Otherwise, instead of a demand for peace, war regulations between states were successful only in that they provided pretexts and slogans for reciprocal accusations. The simple truth is that this has been acknowledged gradually since World War I. Yet, the facade of traditional conceptual inventories still is very strong ideologically. For practical reasons, states have an interest in utilizing so-called classical concepts, also when, in other cases, they are ignored as obsolete and reactionary. Furthermore, since 1900, jurists of European international law deliberately have repressed any recognizable picture of a new reality.[50]

When this is taken for granted with respect to the distinction between traditional European state war and a democratic war of national liberation, then, as Gambetta proclaimed in September 1870, the way is open for an improvised all-out war of national liberation. The 1907 Hague Convention, as did its collective forerunners in the 19th century, sought a compromise with respect to the *Francs-tireurs*. It demanded certain conditions whereby an improvised fighter with an improvised uniform would be recognized as a combatant in the sense of international law: responsible leaders; firm and clearly visible badges of rank; and, above all, open display of weapons. The conceptual obscurity

50. "Without being aware of it, toward the end of the 19th century European international law had lost the consciousness of the spatial structure of its former order. Instead, it had adopted an increasingly more superficial notion of a universalizing process that it naively saw as a victory of European international law. It mistook the removal of Europe from the center of the earth in international law for Europe's rise to the center." See Schmitt, *The* Nomos *of the Earth*, op. cit., p. 233.

of the 1907 Hague Convention and the 1949 Geneva Conventions is a great and complicated problem.[51] The partisan is still the one who refuses to carry weapons openly, who fights from ambush, and who uses the enemy's uniform, as well as true or false insignias and every type of civilian clothing as camouflage. Secrecy and darkness are his strongest weapons, which logically he cannot renounce without losing the space of irregularity, i.e., without ceasing to be a partisan.

The military leadership of the regular Prussian army in no way was based on a lack of intelligence or on ignorance regarding the significance of guerrilla warfare. That one can see in an interesting book written by a typical Prussian General Staff officer, who knew the *Francs-tireurs* war of 1870/71.[52] Colmar Freiherr von der Goltz, the author, died in World War I as the leader of a Turkish army and was called Pasha Goltz. In all objectivity and with great precision, the young Prussian officer recognized the decisive failure of the republican military campaign, and observed: "Gambetta wanted to lead the great war, and lead it he did, but to his misfortune, because a small war, a guerrilla war,

51. The confusion was impenetrable, not only in political propaganda and counter-propaganda (which is its rightful place), and not only in discussion of particularly controversial issues (such as that of the Yugoslav citizen Lazar Vracaric, who in November 1961 was arrested by German authorities in Munich), but unfortunately also in juridical literature as soon as consciousness of the concrete concepts of European international law was lost. That is indicated by Schmid, "Die völkerrectliche Stellung des Partisanen im Kriege," op. cit. In some places, Rentsch is not convinced, and wants to put partisans "under the protection and umbrella of international law," whereby the true partisan would become quite acceptable as a supplementary weapon. See *Partisanenkampf, Erfahrungen und Lehren*, op. cit. Altogether, this is the result of the destruction of the *jus publicum Europaeum* and its humane-rational concepts of war and enemy. The re-barbarizing of the laws of war is a supplementary chapter in the extraordinary book by F. J. P. Veale, *Advance to Barbarism* (Appleton, WI: C. C. Nelson Publishing Company, 1953).

52. Colmar Freiherr von der Goltz, *Léon Gambetta und seine Armeen* (Berlin: F. Schneider, 1877).

would have been much more dangerous for the German army in France at that time."[53]

Although also late, the Prussian-German Supreme Command ultimately did comprehend partisan warfare. On May 6, 1944, the Supreme Command of the German armed forces issued the guidelines for fighting partisans. Thus, before its own end, the German army rightly recognized the partisan. In the meantime, the guidelines of May 1944 also have been recognized as an extraordinary regulation by one of Germany's enemies. After World War II, English Brigadier General Aubrey Dixon, together with Otto Heilbrunn, published a significant book on the partisan, which reprinted extensively the German guiding principles as a typical example of the proper way to fight partisans. English General Sir Reginald F. S. Denning remarks in his foreword to the Dixon-Heilbrunn book that, in his view, mentioning the German partisan regulation of 1944 did not detract from the book, that the guidelines of the German army dealt with the struggle against Russian partisans.[54]

Two events at the German war's end (1944/45) are not attributable to the German armed forces, but rather can be explained as antitheses to them: the German *Volkssturm* [National Storm]

53. "With the further penetration of the invading army, all cadres became weaker and the supplies slower.... That all demonstrated the venturesome irregular troops of the enemy. Nevertheless, Gambetta wanted the great war. The numerical strength of his army and also his combat operations should have been glorious and imposing in order to justify his defense of the nation" (ibid., p. 32). Dr. J. Hadrich (Berlin), to whom I am indebted for the book by Goltz, also made me aware of the fact that the Abyssinians, in their resistance against the Italian army of Mussolini in 1935/36, also were vanquished because, instead of a partisan war, they sought to fight a war with regular troops.

54. Cf. Brigadier C. Aubrey Dixon, O. B. E., and Otto Heilbrunn, *Partisanen: Strategie und Taktik des Guerillakrieges* (Frankfurt a/M-Berlin: Bernard & Graefe Verlag für Wehrwesen, 1956), pp. xiv and 213–40.

and the so-called Werewolf. The *Volkssturm* was called up by an edict on September 25, 1944, as a territorial militia for national defense. During its operations, its members were considered to be soldiers in the sense of national defense regulations and as combatants in the sense of the 1907 Hague Convention. Their organization, outfitting, engagement, fighting spirit, and casualties are described in the recent publication of Major-General Hans Kissel, who since November 1944 was chief of the *Volkssturm*'s operations staff. Kissel reports that in the West the *Volkssturm* was considered by the Allies to be fighting troops, whereas in the East the Russians treated it as a partisan organization and shot the prisoners. Different from this territorial militia, the Werewolf was considered to be a partisan organization of youth. Dixon and Heilbrunn report on the outcome: "Some few budding Werewolves were caught by the Allies, and that ended the matter." The Werewolf was characterized as an "attempt to unleash a children's sniper war."[55] In any case, we need not dwell any further on this.

After World War I, the victors disbanded the German General Staff and forbade its reestablishment, in accord with Art. 160 of the Versailles Treaty of June 28, 1919. It was consistent with the logic of history and international law that the World War II victors (above all, the United States and the Soviet Union), which meanwhile had outlawed war fought as a duel consistent with classical European international law, given their common

55. Hans Kissel, *Der Deutsche Volkssturm 1944/45: Eine territoriale Miliz der Landesverteidigung* (Frankfurt a/M: Verlag E. S. Mittler & Sohn, 1962). The information on the different treatment in the West and the East is on p. 46. The term "children's sniper war" was used by Erich F. Pruck in his review of the Kissel book in *Zeitschrift für Politik*, N. F. 9 (1962), pp. 29ff. Pruck rightly remarks that "the borders between legal combat (in the sense of the 1907 Hague Convention) and partisans is unclear." Cf. Dixon and Heilbrunn, *Partisanen*, op. cit., p. 3, n. 24.

victory over Germany, also outlawed and destroyed the Prussian state. Law No. 46 of the Allied Control Commission of February 25, 1947, reads: "The Prussian state, which for long has been the agency of militarism and reaction in Germany, has *de facto* ceased to exist. Guided by the idea of the maintenance of peace and security of nations, and with the wish to ensure further restoration of political life in Germany on a democratic basis, the Control Commission orders the following: Art. 1 The Prussian state, with its government and all its administrative divisions, is dissolved."

The Partisan as a Prussian Ideal in 1813 and the Turn to Theory

It was no Prussian soldier and no reform-minded Prussian officer of the Prussian General Staff, but rather a Prussian ministerial president, Otto von Bismarck, who, in 1866, to avoid being defeated, "wanted to take every weapon in hand to be able to unleash the national movement not only in Germany, but also in Hungary and Bohemia," against the Hapsburg monarchy and Bonapartist France. Bismarck was determined to get the Acheron moving. He was pleased to use the classical citation *Acheronta movere* [mobilize the netherworld], but, of course, he preferred to blame his internal political opponents. Acherontic plans were far from the minds of both the Prussian King, William I, and the chief of the Prussian General Staff, Helmuth von Moltke; such thinking must have appeared to them to be uncivil and also un-Prussian. The word *acherontic* also would have been considered too strong for the German government and the General Staff with respect to their weak attempts at revolution during World War I. Certainly in this connection, this also was the case with respect to Lenin's journey from Switzerland to Russia in 1917. But everything that the Germans thought and planned at that time about the organization of Lenin's journey has been surpassed and outdistanced so monstrously by historical developments that our thesis about Prussian incompatibility with partisan warfare thereby is refuted, rather than supported.[56]

Nevertheless, once in its history the Prussian soldier-state had an acherontic moment. That was in the winter and spring of 1812/13, when an elite group of General Staff officers sought to unleash and control the forces of national enmity against Napoleon. The German war against Napoleon was no partisan war. One scarcely can call it a national war; rather, as Ernst Forsthoff rightly says, it was only "a legend with political backdrops."[57] It

56. Cf. Otto von Bismarck, *Gedanken und Erinnerungen*, with an afterword by Ernst Friedländer (Stuttgart: Cotta, 1965), Vol. I, Ch. 20 and Vol. III/1, Ch. 10, where the citation *Acheronta movere* serves to paint the devil on the wall. Bismarck pushes the point, for obvious reasons. In reality, Egmont Zechlin, as a modern historian, asserts that Bismarck had assembled "operational Hungarian troops" and generals such as Georg Klapka and István Türr. The officer corps of the Hungarian Legion was composed of the cream of the Hungarian aristocracy. "But Bismarck also did not shrink from accepting the radical socialist Czech revolutionary and friend of Bakunin, Joseph Fric, into headquarters. He brought Colonel Oreskovic in Belgrade and Minister Garasanin, the leading politicians of the South Slav movement, into play, and through Victor Emanuel and also Klapka and Türr, he made contact with the European revolutionary hero Giuseppi Garibaldi." The conservative-reactionary general of the tsar, with whom he was negotiating, telegraphed him that he would prefer to make revolution than to die. By comparison with this national-revolutionary line in Bismarck's politics, the attempts at revolution by the German government and the German General Staff during World War I in Russia, in the Islamic-Israeli world, and in America were weak and "improvised," according to Zechlin, in "Friedensbestrebungen und Revolutionsversuche," in *Das Parlament*, App. 20, 24, and 25 (May and June 1961). In his richly documented book, *Die Revolution in der Politik Bismarcks* (Göttingen: Musterschmidt Verlag, 1957), Gustav Adolf Rein concludes that: "Bismarck highlighted the revolution he had in view in order to reveal its inner weaknesses, and had decided that he would resurrect the old monarchy and give it new life" (p. 13). Unfortunately, the concrete situation of 1866 in Rein's book is not handled well, which would have served to strengthen his thesis.

57. Ernst Forsthoff, *Deutsche Verfassungsgeschichte der Neuzeit*, 2nd ed. (Stuttgart: W. Kohlhammer Verlag, 1961), p. 84. Forsthoff also demonstrates that

did not take long to marshall all the elemental forces in the firm framework of state order and to direct the struggle of the regular German army against the French army. Nevertheless, this short, revolutionary moment has extraordinary significance for the theory of the partisan.

At this point, one will immediately think of the famous masterwork of the science of war, *On War*, by the Prussian General von Clausewitz, and rightly so. But at this time, Clausewitz was still the younger friend of his teachers and masters, Scharnhorst and Gneisenau, and his book was published only after his death, after 1832. However, there is another manifesto of enmity against Napoleon from the spring of 1813, and that is the most astounding document in the whole history of partisan warfare: the Prussian *Landsturm* [army reserve or local militia] edict of April 21, 1813, i.e., the Prussian king's edict published in all forms in the Prussian legal system. The prototype of the Spanish *Reglamento de Partidas y Cuadrillas* of December 28, 1808, which became known as the *Corso Terrestre* in the well-known decree of April 17, 1809, is unmistakable. But this one was not signed (personally) by the monarch.[58] It is astounding to see the name

the notion that the Prussian *Landwehr*—the type of troops that came next in the bourgeois ideal of a militia—was decisive in achieving victory is a legend. "In fact, use of the *Landwehr* at the beginning of the war was very limited. It could not launch an attack, which means its moral energy and military impact were negligible. It was prone to confusion and panic. Only with the greater length of the war, when it had spent more time in battle, did the *Landwehr* also increase its value in combat. Under these circumstances, the claim that the *Landwehr* was decisive in achieving victory belongs in the realm of fable." Ernst Rudolf Huber deals with this time in spring 1813, especially with the *Landsturm* edict, in his *Deutsche Verfassungsgeschichte seit 1789* (Stuttgart: W. Kohlhammer, 1957–), Vol. I, §7, p. 213; and in his *Heer und Staat in der deutschen Geschichte* (Hamburg: Hanseatische Verlagsanstalt, 1938), pp. 144ff.

58. It was issued as a decree of a *Junta Suprema*, because the legitimate monarch then was unavailable. Cf. Fernando Solano Costa, "La Resistencia Popular

of a legitimate king on such a call to partisan warfare. This ten pages of Prussian legal compilation in 1813 certainly is one of the most unusual legal documents in world history.

Every citizen, according to the royal Prussian edict of April 1813, is obligated to resist the invading enemy with weapons of every type. Axes, pitchforks, scythes, and hammers are (in §43) expressly recommended. Every Prussian is obligated to refuse to obey *any* enemy directive, and to injure the enemy with all available means. Also, if the enemy attempts to restore public order, no one should obey, because in so doing one would make the enemy's military operations easier. It is expressly stated that "intemperate, unrestrained mobs" are less dangerous than the situation whereby the enemy is free to make use of his troops. Reprisals and terror are recommended to protect the partisans and to menace the enemy. In short, this document is a Magna Carta for partisan warfare. In three places—in the introduction and in §8 and §52—the Spanish and their guerrilla war are mentioned expressly as the "model and example" to follow. The struggle is justified as self-defense, which "sanctifies all means" (§7), including the unleashing of total disorder.

As I have said, there was no German partisan war against Napoleon. The *Landsturm* edict was changed three months later, on July 17, 1813, and was cleansed of all partisan dangers and of every acherontic dynamic. Everything that followed was played out in the struggles of the regular army, as the dynamic of the national impulse was played out with the regular troops. Thus,

en la Guerra de la Independencia: Los Guerrilleros," op. cit., pp. 415f. The Swiss "Everyman's Guide to Guerrilla Warfare" is no official regulation, but a work issued by the board of directors of the Swiss Non-Commissioned Officers. It would be very illuminating to compare its individual directives (for example, warning the population not to observe the directives of an enemy power) with the express regulations of the Prussian *Landsturm* edict of 1813, in order to bring to mind the similar situation, on the one hand, and the technical and psychological progress, on the other.

Napoleon could gloat that, in the four years of French occupation of German soil, no German civilian took a shot at a French uniform.

Then wherein lies the special significance of this short-lived Prussian decree of 1813? It is the official document of a legitimation of partisans for national defense, and certainly a special legitimation, namely, from a spirit and a philosophy that dominated the Prussian capital, Berlin. The Spanish guerrilla war against Napoleon, the Tyrolean rebellion of 1809, and the Russian partisan war of 1812 were elemental, autochthonous movements of devout Catholic or Orthodox peoples, whose religious traditions were not based on the philosophical spirit of revolutionary France, and to this extent were *under-developed.* The Spaniards, in a furious letter to Napoleon's Governor-General, Louis Nicolas Davout, in Hamburg (dated Dec. 2, 1811), called Napoleon's troops a bunch of assassins, a collection of 300,000 badly-led, superstitious monks, which could not be compared with the diligent, hard-working, intelligent Germans. Berlin in the years 1808–13 was infused with a spirit that was thoroughly consistent with the philosophy of the French Enlightenment, so consistent that it was the equal of it, if not allowed to feel superior to it.

Johann Gottlieb Fichte, a great philosopher; highly-educated and brilliant military men like Scharnhorst, Gneisenau, and Clausewitz; and a poet mentioned earlier, Heinrich von Kleist, who died in November 1811, all recognized the enormous spiritual potential of the effective Prussian intelligentsia at that critical moment. The nationalism of this Berlin intellectual stratum was not just a matter of some simple or even illiterate people, but rather of the educated elite. In such an atmosphere, which united an aroused national feeling with philosophical education, the partisan was discovered philosophically, and his theory became historically possible. A theory of war also was part of this covenant, which is demonstrated by the letter that Clausewitz, as an "unknown military man" in 1809, wrote from Königsberg to Fichte, "the author of an article on Machiavelli." With great respect, the Prussian officer informed the famous philosopher that Machiavelli's theory of war was too dependent on antiquity, and that today one "achieves much more through the continuous revival of individual forces than through aesthetic form." The new weapons and masses that Clausewitz had in mind expressed this principle completely. Ultimately, he said, the courage of the individual facing imminent battle is decisive, "especially in the best of all wars, when a people on its own soil is led to fight for freedom and independence."

The young Clausewitz knew the partisan from the Prussian insurrection plans of 1808–13. In 1810–11, he had given lectures on guerrilla warfare at the General War College in Berlin, and was one of the most important military experts on guerrilla warfare not only in the technical sense, but also in the deployment of light, mobile troops. Guerrilla warfare became for him, as for other reformers of his circle, "above all, a political matter in the highest sense, meaning precisely of a revolutionary character. Acknowledgement of armed civilians, of insurrection, of revolutionary war, resistance, and rebellion against the existing order, even when embodied in a foreign regime of occupation—this was a novelty for Prussia, something 'dangerous,' which similarly fell outside the sphere of lawful states." With these words, Werner Hahlweg came to the core of the matter. Yet, he also added: "The revolutionary war against Napoleon, as the Prussian reformers imagined it, of course did not occur." It was a "half-insurrectional war," as Friedrich Engels called it. Nevertheless, the famous professional report of February 1812 remains significant with respect to the "driving impulses" (Hans Rothfels) of the reformers; with the help of Gneisenau and Hermann de Boyen, Clausewitz had conceived of it before he went to Russia. It is a "document of sober analysis, both politically and in terms of the General Staff," with reference to the experiences of the Spanish guerrilla war, and is content "to return atrocity for atrocity, outrage for outrage." Here, the Prussian *Landsturm* edict of April 1813 is clearly recognizable.[59]

Clausewitz must have been very disappointed that everything he had hoped for from the insurrection had "failed."[60] He always had considered wars of national liberation and partisans ("party followers," as he called them) to be essential parts of "the exploding forces in war," and had worked them systematically into his theory of war. Especially in Book 6 of *On War*, "Defense," and in the famous Chapter 6B of Book 8—"War is an Instrument of Politics"[61]—he also recognized the new "power." Moreover, one finds astoundingly profound comments by him, such as the place regarding civil war in the Vendée: that at times some few individual partisans were able even to "use the name of an army."[62] Yet, he remained a reform-minded professional

59. Hahlweg, *Preussische Reformzeit und revolutionärer Krieg*, op. cit., pp. 54 and 56. Clausewitz's letter to Fichte is published in Fichte's *Staatsphilosophischen Schriften*, ed. Hans Schulz and Reinhard Strecker (Leipzig: F. Meiner, 1925), Supp. Vol. I, pp. 59–65; concerning the "three declarations," see Ernst Engelberg in the introduction to *Vom Kriege* (Berlin: Verlag des Ministeriums für Nationale Verteidigung, 1957), pp. xlvii–l.

60. Letter to Marie von Clausewitz dated May 28, 1813: "By contrast, everything that was expected of support from the people behind enemy lines also appears to have failed. This is the one thing that I did not expect, and I must confess that at moments this memory makes me very sad." Karl Linnebach, *Karl und Marie von Clausewitz: Ein Lebensbild in Briefen und Tagebuchblättern* (Berlin: M. Warneck, 1925), p. 336.

61. Clausewitz, *On War*, op. cit., pp. 357ff. and 605ff., respectively (translation altered).

62. An army is "...all the forces located in a given theater." It "would be sheer pedantry to claim the term 'army' for every band of partisans that operates on its own in a remote part of the country. Still, we must admit that no one thinks it odd to talk of the 'army' of the Vendée during the French Revolutionary Wars, though it was frequently little more than a band of partisans." Clausewitz, *On War*, op. cit., pp. 280–81. Cf. also note 79 (example: Algeria).

officer of a regular army of his time, who could not let the seeds that we see here be developed to their ultimate consequence. As will become evident, that was possible to see only later, and it required an active professional revolutionary. Clausewitz still thought too much in classical categories when he, in the "strange trinity of war," associated the people only with "blind instincts" of hatred and enmity, the commander-in-chief and his army with "courage and talent" as free activities of the soul, and the government as the purest rational handling of war as an instrument of politics.

The moment in which the partisan first entered into a new, decisive role, as a new, formerly unrecognized figure of the world-spirit, was concentrated in this short-lived Prussian *Landsturm* edict of April 1813. It was not the will of resistance of a valiant, bellicose people, but rather education and intelligence that opened this door to the partisan and gave him a legitimation based on philosophy. One could say that he had become philosophically accredited and socially presentable. Until then, that had not been the case. In the 17th century, he was a debased figure in a picaresque novel; in the 18th century, until the time of Maria Theresa and Frederick the Great, he was a pandour and hussar. But now, in Berlin during the years 1808–13, he was discovered and respected not only in a military-technical sense, but also philosophically. At least for a moment, he acquired a historical rank and a spiritual consecration. That was an event that he never could forget.

This is decisive. We have spoken of the theory of the partisan. Now, a political *theory* of the partisan—beyond a technical-military classification—was possible in light of this successful accreditation in Berlin. The spark ignited by the Spanish against the North in 1808 found a theoretical form in Berlin that made it possible to capture the partisan in his glow and to negotiate his existence in other hands.

To begin with, the traditional devoutness of people in Berlin at that time was as little threatened as was the political unity of the king and the people. It even appears to have been strengthened, rather than endangered, by the exorcism and glorification of the partisan. The acheron that had been unleashed was returned immediately to the channels of state order. After the wars of independence, Hegel's philosophy was dominant in Prussia. It sought a systematic mediation of revolution and tradition.[63] It could be considered to be conservative, and it was. But it also conserved the revolutionary sparks, and through its philosophy of history of the ongoing revolution it provided the Jacobins with a dangerous ideological weapon, even more dangerous than Rousseau's philosophy. This historical-philosophical weapon fell into the hands of Karl Marx and Friedrich Engels. Yet, both German revolutionaries were more thinkers than activists of revolutionary war. Only through a Russian revolutionary, through Lenin, did Marxism become a doctrine of world-historical power, which it is today.

From Clausewitz to Lenin

Hans Schomerus, whom we have cited as an expert on the partisan, titles a section of his comments (which he made available to me in manuscript) "From Empecinado to Budjonny." That means: from the partisan of the Spanish guerrilla war against Napoleon to the organizer of the Soviet cavalry, the leader of the Bolshevik war in 1920. Such a heading illuminates an interesting technical-military line of development. However, for us, who have the

63. Joachim Ritter, *Hegel und die französische Revolution* (Cologne and Opladen: Westdeutscher Verlag, 1957). Very illuminating for our context is Reinhart Koselleck's formulation: "The sociological fact is that bourgeois intelligence as a whole and the historical consciousness of Prussian officials are the same phenomenon in that they *find their state in the spirit of the state.*" See Reinhart Koselleck, *Staat und Gesellschaft in Preussen 1815 bis 1848*, in the series titled *Industrielle Welt*, Vol. 1, ed. Werner Conze (Stuttgart: Ernst Klett, 1962), p. 90.

theory of the partisan in view, this consideration directs attention too much to military-technical questions of the tactics and strategy of mobile warfare. We must keep the development of the concept of the political in view, which precisely here takes a subversive turn. In the 18th and 19th centuries, the classical, fixed concept of the political was based on the *state* of European international law, and had bracketed war in classical international law, i.e., had made it purely state war. Since the onset of the 20th century, this state war with its bracketing has been destroyed and replaced by wars of revolutionary parties. For this reason, we use the following heading: "From Clausewitz to Lenin." Obviously, in a sense, therein lies (in contrast to a technical-military limitation) a certain opposite danger, i.e., that we will become lost in historical-philosophical diversions and genealogies.

The partisan here is a surer focal point, because he is kept apart from such general philosophical-historical genealogies and can lead us back to the reality of revolutionary development. Marx and Engels recognized that contemporary revolutionary war is no barricade war in the old style. Engels, in particular, who wrote many military treatises, always emphasized this. But he considered it possible that bourgeois democracy, with the help of universal suffrage, could create a proletarian majority in parliament, and thus legally could turn the bourgeois social order into a classless society. Consequently, one could say that Marx and Engels also called for a completely non-partisan revisionism.

Not so with Lenin, who recognized the inevitability of force and bloody, revolutionary civil war and state war, and thus also approved of partisan warfare as a necessary ingredient of the total revolutionary process. Lenin was the first to fully conceive of the partisan as a significant figure of national and international civil war, and he sought to transform the partisan into an effective instrument of the central Communist Party leadership. As far as I can see, this appeared for the first time in an article titled "The Partisan Struggle," in the Russian periodical *The Proletarian,* dated September 30/October 13, 1906.[64] It is a clear extension of the recognition of enmity and friendship that began in his 1902 pamphlet *What Is To Be Done?*, above all with the turn against Peter Struve's *Objectivism*. Thereby, "the professional revolutionary was consistently in place."[65]

Lenin's article on partisans concerns the tactics of socialist civil war, and is directed against widespread social-democratic opinion at the time that the goal of a proletarian revolution should arise independently as a mass movement in parliamentary countries, meaning that the direct application of force was outdated. For Lenin, partisan warfare was consistent with the methods of civil war and concerned, like everything else, a purely tactical or strategic question of the concrete situation. Partisan warfare, as Lenin said, is "an inevitable form of struggle," which one utilizes without dogmatism or preconceived principles, just as one must utilize other legal or illegal, peaceful or forceful, regular or irregular means according to the situation. The goal is the communist revolution in all countries; what serves this goal is good and just. Consequently, the partisan problem is also very easy to solve: if partisans are controlled by the Communist Central Committee, they are freedom fighters and glorious heroes; if they shun this control, they are anarchistic riffraff and enemies of humanity.

64. W. I. Lenin, *Sämtliche Werke*, Vol. 10, 2nd ed. (Vienna: 1930), pp. 120f. I cite here the German edition of Lenin's military writings, published in East Berlin, 1961: "Vom Krieg, Armee und Militärwissenschaft," Vol. 1, pp. 294–304. It is a remarkable coincidence that George Sorel's "Réflexions sur la Violence" appeared in Paris in the same year (1906), in the periodical *Mouvement Socialiste*. An annotation by Rentsch (*Partisanenkampf: Erfahrungen und Lehren*, op. cit., p. 203n.) led me to the book by Michael Prawdin, *Netschajew—von Moskau verschwiegen* (Frankfurt a/M: Athenäum, 1961), p. 176, where Lenin already in 1905 spoke of the necessity of guerrilla warfare. The genuine text is still to be verified.

65. Peter Scheibert, "Über Lenins Anfänge," in *Historische Zeitschrift* 182 (1956), p. 564.

Lenin was a great expert on and admirer of Clausewitz. He had studied *On War* intensively during World War I (1915), and in his notebooks he copied quotations in German and made comments in Russian with underscoring and exclamation marks. In this way, he created one of the most remarkable documents of world history and intellectual history. From a fundamental consideration of these quotations, marginalia, underscoring, and exclamation points,[66] it is possible to develop the new theory of absolute war and absolute enmity that has determined the age of revolutionary war and the methods of modern cold war. What Lenin was able to learn from Clausewitz, and what he learned painstakingly, was not only the famous formula of war as the continuation of politics. It was the further recognition that the distinction of friend and enemy in the age of revolution is primary, and that it determines war as well as politics. For Lenin,

66. A German edition of Lenin's *Tetradka* [notebooks] on Clausewitz's *On War* was published in Berlin in 1957 by the Institute of Marxism-Leninism of the Socialist Unity Party (a.k.a. East German Communist Party). The most extensive and important presentation and analysis of the *Tetradka* is by Werner Hahlweg in an article titled "Lenin und Clausewitz," in *Archiv für Kulturgeschichte*, Vol. 36 (1954), pp. 30–39 and 357–87. Hahlweg is also the editor of the latest edition of *Vom Kriege*, op. cit. According to Hahlweg, Lenin's original contribution was that he extended Clausewitz from the stage of the (at first, bourgeois) Revolution of 1789 to the proletarian revolution of 1917, and recognized that war had left the ranks of national and state wars to become a class war at the place where Marx and Engels hoped for an economic crisis. With help from the formula "war is the continuation of politics," Lenin declared "now the whole core question of revolution is in its struggle: recognition of the essence (class analysis) of the world war, together with attendant problems such as opportunism, defense of the fatherland, the fight for national liberation, the distinction between just war and unjust war, the relation between war and peace, revolution and war, the end of imperialist wars through the internal revolt of the working class, revision of the Bolshevik Party program" (Hahlweg, ibid., p. 374). It appears to me that every point that Hahlweg rightly makes here provides a touchstone for the enemy concept.

only revolutionary war is genuine war, because it arises from absolute enmity. Everything else is conventional play.

Lenin stressed the distinction between war (*Voina*) and play (*Igrá*) in a marginal note in Chapter 23, Book II ("Key to the Land"). Logically, this was the decisive step in the destruction of the bracketing [of war] between states in continental European international law that had been achieved in the 18th century. The Congress of Vienna was so successful in restoring [the bracketing of war after the French Revolution] that it lasted into World War I. Not even Clausewitz ever imagined that it would be destroyed. By comparison with a war of absolute enmity, the bracketed war of classical European international law, recognizing accepted rules, is similar to a duel between cavaliers seeking satisfaction. To a communist like Lenin, inspired by absolute enmity, such a type of war must have appeared to be mere play, which he might join in if the situation demanded, but which basically he would find contemptible and ludicrous.

The war of absolute enmity knows no bracketing. The consistent fulfillment of absolute enmity provides its own meaning and justification. The question is only: Is there an absolute enemy, and if so who is he? For Lenin, the answer was obvious, and the fact that he made absolute enmity serious made him superior to all other socialists and Marxists. His concrete absolute enemy was the class enemy—the bourgeois, the essential capitalist, and the social order in countries where this bourgeois capitalist was dominant. Cognizance of the enemy was the secret of Lenin's enormous effectiveness. His understanding of the partisan was based on the fact that the modern partisan had become the true irregular and, thereby, the strongest negation of the existing capitalist order; he was called to be the true executor of enmity.

Today, the irregularity of the partisan consists not only in a military "line," which was the case in the 18th century, when he was only a "light troop," and not only in the pride of wearing the uniform of a regular troop. The irregularity of the class struggle challenges not only a line, but the whole structure of political and social order. This new reality was conceived with philosophical consciousness by the Russian revolutionary Lenin, and the alliance of philosophy and the partisan that he forged unleashed unexpected new and explosive forces. It caused nothing less than the destruction of the whole Eurocentric world that Napoleon had hoped to rescue and that the Congress of Vienna had hoped to restore.

The bracketing of *inter*state regular war and the overcoming of *intra*state civil war had become so accepted in 18th century Europe that intelligent men of the *ancien régime* also could not conceive of the destruction of this type of regularity, not even after the experiences of the French revolutions of 1789 and 1793. For such, they found only the language of a general horror, and made basically incongruous, childish comparisons. A great and courageous thinker of the *ancien régime*, Joseph de Maistre, had foreseen brilliantly what was at stake. In a letter written in the summer of 1811,[67] he declared that Russia was ripe for revolution, yet hoped that it would be, as he put it, a *natural* revolution, not an enlightened-European revolution like the French. What he feared most was an *academic* Pugachev. Thus, he took pains to make clear what he considered to be the real danger, namely, an alliance of philosophy with the elemental forces of an insurrection. Who was Pugachev? He was the leader of a peasant and Cossack rebellion against Catherine II, who put a price on his head. He was executed in 1775. An *academic* Pugachev would

67. *Europa und Russland: Texte zum Problem des westeuropäischen und russischen Selbstverständnisses*, ed. Dmitrij Tschižewskij and Dieter Groh (Darmstadt: Wissenschaftliche Buchgesellschaft, 1959), p. 61. Letter to Rossi dated August 15, 1811. On Maistre's critique and prognoses of Russia, see Dieter Groh, *Russland und das Selbstverständnis Europas: Ein Beitrag zur europäischen Geistesgeschichte* (Neuwied: Herrmann Luchterhand Verlag, 1961), esp. pp. 105ff. The book is also of great significance for our theme in many other respects.

be a Russian who "started a European-style revolution." That would produce a series of horrible wars, and if they went too far, "I would not have the words to tell you what one would then have to fear."

The vision of smart aristocrats is astounding, as much as in what they saw, namely, the possibility and danger of an alliance of Western intellect and Russian rebellion, as in what they did not see. With their timely and orderly dates—St. Petersburg in the summer of 1811—they found themselves to be in the closest proximity to the Prussian army reformers. Yet, given their own nearness to the reform-minded professional officers of the Prussian General Staff, they did not notice the intensive contacts that these officers still maintained with the imperial court in St. Petersburg. They knew nothing of Scharnhorst, Gneisenau, and Clausewitz, and they failed to see the fatal flaw in linking their names with Pugachev. The profundity of a significant vision was lost, and what remained was *bonmot* [repartee] in the style of Voltaire or even Antoine de Rivarol. If one still thinks in terms of the alliance between Hegel's philosophy of history and unchained mass forces, such as the Marxist professional revolutionary Lenin forged, then the formulation of the brilliant Maistre would shrink to a small verbal effect in rooms or anterooms of the *ancien régime*. The language and conceptual world of bracketed war and prescribed enmity no longer were any match for absolute enmity.

From Lenin to Mao Tse-tung

In the assessment of experts, during World War II Russian partisans diverted approximately twenty German divisions, thereby contributing essentially to the war's outcome. In his book on the Great Fatherland War (1941–45), the official Soviet historiographer, Boris Semenovitch Telpuchovski, describes the glorious partisans who wrecked havoc behind enemy lines. In the enormous spaces of Russia, with seemingly endless fronts thousands of kilometers long, every division in the German war effort was irreplaceable. Stalin's fundamental concept of partisans was that they must fight behind enemy lines, consistent with the maxim: in the rear, partisans; at the front, fraternization.

Stalin succeeded in linking the strong potential of national and homeland resistance—the essentially defensive, telluric power of patriotic self-defense against a foreign invader—with the aggressivity of the international communist world revolution. The linking of these two heterogeneous movements dominates contemporary partisan warfare around the world. Consequently, until now a communist element has been in the forefront with its goal-oriented policy and its dependence on Moscow or Peking. The Polish partisans, who during World War II fought against the Germans, were sacrificed by Stalin in a gruesome way. The partisan struggle in Yugoslavia (1941–45) was not only a common national defense against the foreign invader, but also a very brutal internal struggle between communist and monarchist partisans. In this fraternal struggle, the Communist Party leader Tito, with the help of Stalin and England, conquered and destroyed his internal Yugoslav enemy, namely General Drazha Mihailovich, who originally had been supported by England.

The greatest practitioner of contemporary revolutionary war became as well its most famous theoretician: Mao Tse-tung. Some of his writings are "today required reading at Western war colleges" (Hans Henle). Since 1927, he had assembled experiences in communist actions and then used the 1932 Japanese invasion to develop systematically all modern methods of national and international civil war. The "long march" from southern China to the Mongol border began in November 1934. It covered approximately twelve thousand kilometers and suffered enormous casualties. But it constituted a series of partisan achievements and experiences that culminated in the Chinese Communist

Party becoming a peasant- and soldier-party with partisans at the core. It is a significant coincidence that Mao completed his most important writings in 1936–38, i.e., the same years in which Spain defended itself against a war of national liberation sponsored by the international communist movement. But in the Spanish Civil War, partisans played no significant role, whereas Mao credited the victory over his national opponent, Chiang Kai-shek, totally to the experiences of the Chinese partisan war against the Japanese and the Kuomintang.

For our theme, Mao's most important formulations are found in a 1938 work titled "Problems of Strategy in Guerrilla War Against Japan."[68] Yet, other of his writings also must be mentioned in order to make this new Clausewitzian theory of war completely understandable.[69] In fact, his theory of war is a consistent and systematic continuation of Prussian General Staff officers' concepts. Clausewitz, the contemporary of Napoleon I, could not have conceived of the degree of totality that today is obvious in the revolutionary war of the Chinese communists. The characteristic picture Mao provides is found in the following comparison: "In our war, the armed people and the small partisan war on one side, and the Red Army on the other, is comparable to the two arms of a man or, to put it more practically: the morale of the people is the morale of the nation in arms. For this reason, the enemy is afraid."

The "nation in arms": as is well-known, this was also the slogan of the professional officers of the Prussian General Staff,

68. See *Selected Works of Mao Tse-Tung*, Vol. II (Peking: Foreign Languages Press, 1965), pp. 79–112.

69. Cf. Theodor Arnold, *Der revolutionäre Krieg* (Pfaffenhof/Ilm: Ilmgau Verlag, 1961), pp. 22f. and 97ff.; Hellmuth Rentsch, *Partisanenkampf: Erfahrungen und Lehren* (Frankfurt a/M: Bernard & Graefe Verlag für Wehrwesen, 1961), especially pp. 150–201 (the example of China); Klaus Mehnert, *Peking und Moskau* (Stuttgart: Deutsche Verlags-Anstalt, 1962), p. 567; and Hans Henle, *Mao, China und die Welt von heute* (Stuttgart: Union, 1961).

who organized the war against Napoleon. Clausewitz was one of them. We have seen that the strong national energies of a particular educational level of the regular army already had been garnered at that time. Also, the most radical contemporary military thinkers also distinguished between war and peace, and considered war to be a clearly separate state of exception from peace. As a professional officer of a regular army, Clausewitz could not think through the logic of the partisan as systematically as could professional revolutionaries like Lenin and Mao. However, with Mao there is still a concrete factor with reference to the partisan, whereby he came closer than Lenin to the core of the matter, which made it possible for him to think the partisan through to the end. In short: Mao's revolution was more tellurically based than was Lenin's.

There are great differences between the Bolshevik *avant-garde*, which seized power in Russia under Lenin's leadership in October 1917, and the Chinese Communists, who in 1949 won in China after more than 20 years of war—differences in their internal organizational structure, as well as in their relation to the land and the people they conquered. In view of the enormous reality determined by a telluric partisan, the ideological controversy concerning whether Mao taught a true Marxism or Leninism becomes as secondary as the question of whether old Chinese philosophers also made remarks similar to those of Mao. This question deals with a concrete "red elite" characterized by the partisan struggle. Most essential, as Ruth Fischer has formulated so clearly, is that the Russian Bolsheviks of 1917, from a national standpoint, were a minority "led by a group of theoreticians whose majority were émigrés." In 1949, the Chinese Communists under Mao and his friends had struggled for two decades on their own national soil with a national opponent (the Kuomintang) in an enormous partisan war. It may have been that their provenance was the urban proletariat, similar to the

Russian Bolsheviks who hailed from St. Petersburg and Moscow, but once the Chinese Communists came to power they brought with them the characteristic experiences of the most difficult defeats, together with the organizational competence "to plant [their core principles] in a peasant milieu, and there to develop them further in a new and unforeseen way."[70]

Here lies the deepest origin of the "ideological" differences between Soviet Russian Communism and Chinese Communism. But it also can be traced to an inner contradiction in the situation of Mao, who linked a universal, absolute, global enemy lacking any territorial space—the Marxist class enemy—with a territorially limited, real enemy of the Chinese-Asiatic offensive against capitalist colonialism. It is the antithesis of *One World*, i.e., a political unity of the earth and its humanity, and a plurality of *Grossräumen* [large spatial-political spheres], which are rationally balanced internally and in relation to one another. In a

70. Ruth Fischer, *Von Lenin zu Mao: Kommunismus in der Bandung-Aera* (Düsseldorf-Cologne: Eugen Diederichs Verlag, 1956), p. 155. Cf. on the China example and the peasant problem, Rentsch, *Partisanenkampf: Erfahrungen und Lehren*, op. cit., pp. 154f. On the proletariat and the peasantry, see Mehnert, *Peking und Moskau*, ibid., pp. 179ff. See Henle, *Mao, China und die Welt von heute*, ibid.: on the significance of partisan warfare, p. 102; on the red elite, pp. 150ff.; on the special Chinese line of socialism and communism, pp. 161ff. Walt W. Rostow does not deal with the decisive theme of Chinese partisanship, although he well observes the traditional character of the Chinese elite in *The Prospects for Communist China* (New York and London: The Center for International Studies, Massachusetts Institute of Technology, 1954), pp. 10f., 19–21, and 136. He remarks that "Peking's leaders have a strong sense of history," p. 312. He also remarks that the manner of thinking of Chinese communism since Mao's rise to power has been characterized by *mixed political terms*. If this formulation is meant derogatorily, which is conceivable, it nevertheless blocks his view of the core of the matter, namely the question of partisanship and the real enemy. On the controversy concerning the legend of Mao (Benjamin Schwarz and Karl A. Wittfogel), cf. Mehnert, *Peking und Moskau*, ibid., p. 566, n. 12.

poem titled *Kunlun*, Mao depicted the pluralistic image of a new *nomos* of the earth:

> If I could stand above the heavens,
> I would draw my sword
> And cut you in three parts:
> One piece for Europe,
> One piece for America,
> One piece left for China.
> Then peace would rule the world.[71]

In the concrete situation, Mao encountered various types of enmity, which intensified into absolute enmity: racial enmity against the white, colonial exploiter; class enmity against the capitalist bourgeoisie; national enmity against the Japanese intruder; and the growing enmity against his own national brothers in long, bitter civil wars. All of this did not paralyze or relativize enmity, as one might have thought, but rather intensified and strengthened it in a concrete situation. During World War II, Stalin successfully linked the telluric partisan of the national homeland with the class enmity of international communism. In this respect, Mao was many years ahead of Stalin. Also in his theoretical consciousness, Mao took the formula of war as the continuation of politics even further than did Lenin.

Mao's approach is as simple as it is effective. War has its meaning in enmity. Because it is the continuation of politics, it also encompasses politics, at least the possibility that there is always an element of enmity; and if peace contains the possibility of war (which, from experience, unfortunately is the case), it

71. [Tr. Kunlun is a mountain which, as Mao Tse-tung writes, rises far above the earth and into the heavens. It has "seen the best in the world of men." In conclusion, he writes: "Now I say to Kunlun: Neither all your height, nor all your snow is needed." Then follow the lines in the text. Cf. Mao Tse-tung, *Poems* (Peking: Foreign Languages Press, 1976), pp. 20–21. I have altered this official translation, which is inadequate.]

also contains a factor of potential enmity. The question is only whether enmity can be bracketed and regulated, i.e., whether it is relative or absolute. That can be decided only by the belligerents at their own risk. For Mao, who thinks as a partisan, peace today is only a manifestation of real enmity. Enmity also does not cease in so-called cold war, which is not half war and half peace, but rather a situation of enmity with other than open violent means. Besides, only weaklings and illusionists are able to be deceived.

Practically speaking, there is thus the question of what is the quantitative relation between the actions of a regular army during hostilities and other methods of the class struggle that are not overtly military. Here, Mao found a clear formula: revolutionary war is 1/10 overt, regular war and 9/10 not. Based on it, a German General, Helmut Staedke, formulated a definition: a partisan is a fighter who pursues war for the 9/10 and leaves 1/10 to the regular troops.[72] Mao certainly did not overlook the fact that this 1/10 is decisive for the end of the war. Yet, as Europeans of the old tradition, one certainly must emphasize that, when one speaks of war and peace, one is referring to the conventional, classical concepts of war and peace of European bracketed war in the 19th century; thus, not to an absolute, but only to a relative and bracketed enmity.

The regular Red Army appears then only when the situation is mature enough for a communist regime. Only then does the land become openly occupied militarily. Of course, this does not have reference to a peace treaty in the sense of classical

72. Helmut Staedke, in a speech on October 17, 1956, before the Arbeitsgemeinschaft für Wehrforschung [Committee for Weapons Research]. Especially well-known in Germany is J. Hogard, "Theorie des Aufstandskrieges," in *Wehrkunde*, Vol. 4 (October 1957), pp. 533–38; cf. also Colonel Charles Lacheroy, *Action Viet-Minh et Communiste en Indochin ou une leçon de "guerre révolutionaire"* (Paris: Section de Documentation Militaire, 1955), and Arnold, *Der revolutionäre Krieg*, op. cit., pp. 171ff.

international law. The practical significance of such a doctrine was demonstrated most vividly after 1945 with the division of Germany. On May 8, 1945, the military war against defeated Germany ended; Germany then had surrendered unconditionally. Until today, 1963, there is still no formally concluded peace between Germany and the allied victors; until today, the borders between East and West are precisely those lines that American and Soviet regular troops drew as their zones of occupation 18 years ago.

Both the relation (with the 9:1 ratio) of cold war and open hostilities, as well as the deeper, world-political, symptomatic significance of the division of Germany after 1945 are for us only examples to clarify Mao's political theory. Its core lies in the partisan, whose essential characteristic today is true enmity. Lenin's Bolshevik theory recognized and acknowledged partisans. Yet, by comparison with the concrete telluric reality of the Chinese partisans, there was something abstractly intellectual in Lenin's determination of the enemy. The ideological conflict between Moscow and Peking, which has become increasingly stronger since 1962, has its deepest roots in this concrete and dissimilar reality of the true partisan. Also here, the theory of the partisan proves to be the key to knowledge of political reality.

From Mao Tse-tung to Raoul Salan

French professional officers returning from Asia to Europe have spread Mao Tse-tung's fame as the most modern teacher of the conduct of war. In Indochina, old-style colonial war merged with contemporary revolutionary war. There, French professional officers experienced the effectiveness of well thought out methods of subversive warfare and psychological mass terror, and learned that they could be combined with partisan warfare as if they were one. From their experiences, they developed a doctrine of psychological, subversive, and insurrectional warfare, about which there is already a voluminous literature.[73]

Therein, one can recognize the typical product of a manner of thinking characteristic of professional officers, specifically of colonels. No more need be said here about this association with colonels, although it might be interesting to pose the question of whether, on the whole, a figure like Clausewitz is closer to the intellectual type of a colonel than of a general. For us, this question deals with the theory of the partisan and its consistent development. In recent years, a clear and concrete case can be embodied in a general, rather than in a colonel, namely, in the fate of General Raoul Salan. More than other generals, such as Edmund Jouhaud, Maurice Challe, or André Zeller, he is for us the most important figure in this connection. An existential conflict is revealed in the unfolding position of this general, which is the decisive conflict for an understanding of the partisan problem, i.e., when regular troops are fighting against a fundamentally revolutionary and irregular foe not only occasionally, but continuously in a war aimed directly at them.

Salan became acquainted with the colonial war in [French] Indochina as a young officer. During World War II, he was made a member of the Colonial General Staff, and served in this capacity in [French West] Africa. In 1948, he became the commandant of French troops in Indochina. In 1951, he was named

73. I refer summarily to the literature cited in books by Arnold, *Der revolutionäre Krieg*, op. cit.; Rentsch, *Partisanenkampf: Erfahrungen und Lehren*, op. cit.; Raymond Aron, *Paix et Guerre entre les Nations* (Paris: Callmann-Lévy, 1962); Arias, *La Guerra Moderna y la Organisacion Internacional*, op. cit. In addition, see Luis García Arias, *Etudes des Phénomènes de la Guerre psychologique des Ecole Militaire d'Administration de Montpellier* (1959), especially Vol. 2, *Les Formes Nouvelles de la Guerre*, as well as Jacques Fauvet and Jean Planchais, *La Fronde des Généraux* (Paris: Arthaud, 1961); Claude Paillat, *Dossier Secret de l'Algérie* (Paris: Presses de la Cité, 1962); and Peter Paret and John W. Shy, *Guerrillas in the 1960s* (New York: Praeger, 1962), p. 88.

High Commissioner of the French Republic in North Vietnam. In 1954, he led the inquiry into the defeat at Dien-Bien Phu. In November 1958, he was named Supreme Commandant of French armed forces in Algeria. Until then, he could be described politically as on the Left, and yet, in January 1957, a secret organization that might be characterized as a "kangaroo court" made an attempt on his life. But the lessons of war in Indochina and the experiences of the Algerian partisan war were such that he had absorbed the inexorable logic of partisan warfare. The premier of the former Paris government, Pierre Pflimlin, gave him full powers. But on May 15, 1958, at the decisive moment, he helped General Charles de Gaulle gain power by shouting "Vive de Gaulle!" at a public meeting in the forum in Algiers. Yet, he soon became bitterly disappointed in his expectations that General de Gaulle would defend unconditionally France's territorial sovereignty over Algeria, which was guaranteed in the constitution.

Open enmity against General de Gaulle began in 1960. In January 1961, a few of Salan's friends founded a Secret Army Organization (the OAS: Organisation d'Armée Secrète). Salan became its leader when, on April 23, he was called to join the officers' putsch in Algeria. When this putsch began to crumble on April 25, the OAS pursued systematic terrorist actions—systematic in the sense of so-called psychological warfare of modern mass terror—against both the Algerian enemy and the Algerian civilian population, as well as against the civilian population in France. The decisive blow against these terrorist actions occurred in April 1962, when Salan was arrested by the French police. The trial before the highest military court in Paris began on May 15 and ended on May 23, 1962. The indictment specified an attempted forceful overthrow of the legal regime and the terrorist acts of the OAS between April 1961 and April 1962. Since the court allowed for extenuating circumstances, Salan did not get the death penalty, but life in prison (*détention criminelle à perpétuité*).

I have reminded the reader of a few important dates, but there is still no history of Salan and the OAS,[74] and it is not my intention to meddle in the deep internal conflict of the French nation by expressing opinions and making judgments. Here, we can elaborate only a few aspects from the material that has been published, in order to illuminate our substantive question.[75] Many parallels with respect to the partisan come to mind. We will return to one of them on purely heuristic grounds and with all due caution. The analogy between the Spanish guerilla war experienced by the Prussian General Staff (1808–13) and the partisan warfare in Indochina and Algeria experienced by the French General Staff (1950–60) is striking. But so are the great differences, which require no further comment. There is a relationship in the core situations and in many individual fates. Yet, this should not be exaggerated abstractly, as if these situations and fates can be identified with the theories and constructions of all the defeated armies of world history. That would be ridiculous. The case of Prussian General Ludendorff also is in many respects different from that of the Left-Republican Salan. We are concerned only with a clarification of the theory of the partisan.

During the trial before the High Military Court, Salan remained silent. At the beginning of the trial, he provided a long explanation, which began with the statement: *Je suis le chef de*

74. [Tr. In the meantime, several works by and about Salan have appeared, among them: Fabrice Laroche, *Salan devant l'opinion* (Paris: Éditions Saint-Just, 1963); André Figueras, *Raoul Salan, ex-général* (Paris: La Table ronde, 1965); Raoul Albin Louis Salan, *Lettres de prison*, ed. André Figueras (Paris: La Table ronde, 1965); Raoul Albin Louis Salan, *Mémoires* (Paris: Presses de la Cité, 1970); Raoul Salan, *Indochine rouge: le message d'Hô Chi Minh* (Paris: Presses de la Cité, 1975).]

75. *Le Procès de Raoul Salan, compte-rendu sténographique*, in *Le grands procès contemporains*, ed. Maurice Garçon (Paris: Edition Albin Michel, 1962).

l'OAS. Ma responsabilité est donc entière. [I am head of the OAS. Therefore, the responsibility is entirely mine.] In his explanation, Salan refused to call the witnesses he had named (among them, President de Gaulle) and insisted that the trial be limited to the time from April 1961 (the officers' putsch in Algeria) to April 1962 (Salan's imprisonment), whereby his essential motives were obscured and great historical events were isolated, which reduced the facts of the case to those of a normal penal code. He described the violent acts of the OAS as merely responses to the most odious atrocities that can be done to men who did not want to leave their country and did not want their country to be taken away from them. His explanation ended with the words: "I owe an explanation only to those who suffered and died believing a broken promise and fulfilling a betrayed duty. Henceforth, I will remain silent."

Salan actually maintained his silence throughout the whole trial, even during the harsh and insistent questioning of the public prosecutor, who declared that Salan's silence was just a tactic. Finally, after the public prosecutor remarked on the "illogic" of Salan's silence, the president of the High Military Court said that even if this behavior could not be respected, it nevertheless would be tolerated and not treated as contempt of court. When the trial was over, the president asked Salan if he had anything to say in his defense. He answered: "I will open my mouth only to say *Vive la France*!, and to the prosecuting attorney I say simply, *que Dieu me garde*! [May God protect me!]"[76]

76. Five times the defense attorney affirmed the "great silence" of the defendant regarding questions from the public prosecutor (cf. *Le Procès de Raoul Salan*, op. cit., pp. 108 and 157). Salan's reiteration of his declaration that he would remain silent cannot be seen as a disruption of his silence (ibid., pp. 89, 152, and 157), any more than his thanks to the earlier president, Coty, after his depositions (ibid., p. 170). The unusual concluding sentence of the defense attorney's speech is not understandable without Salan's concluding statement, which is found in the transcript of the trial (ibid., p. 480).

The first part of Salan's concluding remark was directed to the president of the High Military Court, and had in view the presumed death sentence. In this situation—at the moment of his execution—he would shout: *Vive la France*! The second part was directed to the prosecuting attorney, and sounded somewhat cryptic. However, it is perfectly understandable that the public prosecutor suddenly became religious, although it was certainly unusual in an all but secular state. Not only did he characterize Salan's silence as arrogant and lacking in remorse, and as an attempt to plead extenuating circumstances for a milder sentence; suddenly, he spoke, as he expressly stated, as "a Christian to a Christian": *un chrétien qui s'adresse à un chrétien*, and told the defendant that he had forfeited the grace of God and incurred eternal damnation. To which Salan answered: *que Dieu me garde*!

One sees the abyss over which the sagacity and rhetoric of a political trial was played out. Yet, for us, it is not the problem of political justice that is interesting,[77] but rather the illumination of a complex of questions that have been thrown into utter confusion by such slogans as total war, psychological war, subversive war, insurrectional war, and covert war, and have occluded the problem of the modern partisan.

The war in Indochina (1946–54) was the "ideal example of a fully-developed modern revolutionary war."[78] Salan had become acquainted with modern partisan warfare in the forests, jungles, and rice fields of Indochina. He had learned firsthand that Indo-Chinese rice peasants could put a battalion of first-rate French soldiers on the run. He saw the squalor of refugees

77. Concerning the extent to which the means and methods of a juridical trial change their object, see "Das Reichsgericht als Hüter der Verfassung" (1929), gloss 5, in Schmitt, *Verfassungsrechtliche Aufsätze*, op. cit., p. 109.

78. Arnold, *Der revolutionäre Krieg*, op. cit., p. 186.

and became acquainted with Ho Chi-minh's underground organization, which the legal French administration overlooked and overplayed. With the exactitude and precision of a member of the General Staff, he observed and scrutinized the new, more or less terrorist conduct of war. At the same time, it occurred to him that what he and his comrades called "psychological" warfare was, together with military-technical action, part of modern warfare. Here, Salan readily could adopt Mao Tse-tung's system of thought; yet, it is well-known that he also had studied the literature concerning the Spanish guerrilla war against Napoleon. In Algeria, Salan faced a situation where 400,000 well-armed French soldiers fought against 20,000 Algerian partisans, with the result that France renounced its sovereignty over Algeria. The Algerian population's loss of human life was 10–20 times greater than on the French side, but the material expenses of the French were 10–20 times higher than those of the Algerians. In short, with his whole existence as a Frenchman and a soldier, Salan was faced with an *étrange paradoxe* [strange paradox] and an *Irrsinnslogik* [insane logic], which could embitter a brave and intelligent man and drive him to attempt a counteroffensive.[79]

79. Raymond Aron speaks of an *étrange paradoxe* of the Algerian situation, in the chapter titled "Determinants et Nombre," in his great work, *Paix et Guerre entre les nations* (Paris: Calmann-Lévy, 1962), p. 245. The term *Irrsinnslogik* is found in the partisan story by Jan Coster, *Der Wächter an der Grenze* (Tübingen: Furche Verlag, 1948). [Tr. Schmitt incorrectly cites Hans Schomerus as the author of this story.]

Aspects and Concepts of the Last Stage

In the labyrinth of such a typical situation for modern partisan warfare, we would like to distinguish four different aspects in order to gain a few clear concepts: the spatial aspect; destruction of social structures; the interlocking global-political context; and finally, the technical-industrial aspect. This sequence is relatively fluid. In concrete reality, these four aspects obviously cannot be isolated as independent spheres; on the contrary, only their intensive reciprocal actions, their mutually functional dependencies can provide the total picture. Discussion of one always contains references to and implications for the other three aspects, and ultimately all flow into the force-field of technical-industrial development.

The Spatial Aspect

Completely independent of the good or ill will of men, of peaceful or hostile purposes and goals, any enhancement of human technology produces new horizons and unforeseeable changes in traditional spatial structures. That is true not only for the external and conspicuous expansions of cosmic spatial exploration, but also for our old, terrestrial living spaces, work spaces, cultural spaces, and even personal spaces. Today, in the age of electric lights, long-range fuel supplies, telephones, radios, and television, the expression "the home is inviolable" produces a completely different type of bracketing from what existed in the age of King John and the Magna Carta of 1215, when the lord of the manor could raise the drawbridge. The technical enhancement of human effectivity shatters whole normative systems, as did the law of the sea in the 19th century. From the bottom of the sea, which has no lord, arose the space that lay before the coast, the so-called continental shelf, as a new sphere of human action. Bunkers for atomic waste were created in the deep of the Atlantic Ocean, which has no lord. Together with spatial structures, technical-industrial progress also changes spatial orders. Law is the unity of order and orientation, and the problem of the partisan is the problem of relations between regular and irregular struggle.

A modern soldier is personally either optimistic or pessimistic about the future. For our problem, that is not so important. In terms of weapons technology, every member of the General Staff thinks practically and purposefully. Consequently, the spatial aspect of war is of theoretical concern to him. The structural variety of so-called theaters of war on land and on sea is an old theme. Since World War I, airspace has become a new dimension, whereby the spatial structure of traditional *theaters* of land and sea were changed.[80] In partisan warfare, a new, complicated, and structured sphere of action is created, because the partisan does not fight on an open battlefield, and does not fight on the same level of open fronts. He forces his enemy into another space. In other words, he displaces the space of regular, conventional theaters of war to a different, darker dimension—a dimension of the abyss,[81] in which the proudly-worn uniform [of the conventional

80. Cf. "The Spatial Perspective of the Theater of War on Land and on Sea" and "Transformation of the Spatial Perspective of Theaters of War," in Schmitt, *The* Nomos *of the Earth*, op. cit., pp. 309–13 and 313–316, respectively, as well as the Berlin Dissertation of Ferdinand Friedensburg, *Der Kriegsschauplatz* (1944).

81. In Dixon and Heilbrunn's book, *Partisanen: Strategie und Taktik des Guerillakrieges*, op. cit., p. 199, partisan warfare is characterized as a struggle "in the abyss of the enemy front, which clearly is not consistent with the general spatial problem of land war and sea war." On this general spatial problem, see my publication titled *Land und Meer: Eine weltgeschichtliche Betrachtung*

soldier] becomes a deadly target. In this way, the partisan on land has an unexpected, but no less effective analogy to a submarine at sea, which opened up an unexpected deep dimension beneath the surface on which sea war in the old style was fought. From an underground lair, the partisan disturbs the conventional, regular play of forces on the open stage. From his irregularity, he changes the dimensions of regular armies not only tactically, but strategically as well. By exploiting their knowledge of the terrain, relatively small partisan groups can tie down great masses of regular troops. We have referred to the "paradox" with the example of Algeria. Clausewitz clearly recognized this and profoundly endorsed it, in that he said that a few partisans who dominate a given terrain can claim the right to be called "an army."

It serves the concrete clarity of the concept that we hold to the telluric-terrestrial character of the partisan, and do not characterize or even define him as a corsair on land. The irregularity of the pirate lacks any relation to regularity. By contrast, the corsair takes booty at sea, and is equipped with a "letter" from a state government; his type of irregularity thus has some relation to regularity, which is why until the Paris Peace of 1856 he was juridically a recognized figure of European international law. To this extent, both the corsair of sea war and the partisan of land war could be compared with each other. A strong similarity and even equality exists above all in the fact that the statement "with a partisan, one fights like a partisan" and the statement "*à corsaire corsaire et demi*" [with a corsair, one fights like a corsair and a half] are saying essentially the same thing. However, the contemporary partisan is something different from a corsair of land war. The elemental antithesis of land and sea remains too great. It could be that the traditional varieties of war, enmity, and booty, which until now have been the basis of the antithesis of land and

[1942] (Cologne-Lövenich: "Hohenheim" Verlag, 1981), and also my book *The* Nomos *of the Earth*, op. cit., pp. 172ff.

sea in international law, one day simply will be dissolved in the crucible of industrial-technical progress. Until now, the partisan always has been a part of the true earth; he is the last sentinel of the earth as a not yet completely destroyed element of world history.

The Spanish guerrilla war against Napoleon came to full light only in the great spatial aspect of this antithesis of land and sea. England supported the Spanish partisans. A maritime power utilized the irregular fighters of land war for its great belligerent undertakings in order to vanquish its continental enemy. Ultimately, Napoleon was defeated not by England, but by the land powers Spain, Russia, Prussia, and Austria. The irregular, typically telluric type of partisan fighting entered into the service of a typically maritime world politics, which relentlessly disqualified and criminalized any irregularity on the sea and in sea war law. Different types of irregularity are concretized in the antithesis of land and sea. When we keep in mind the concrete particularity of the spatial aspects characteristic of *land* and *sea* in the specific forms of their conceptual construction, the analogies are permitted and fruitful. That is especially true of the analogy introduced here for an understanding of the spatial aspect, namely that of how the sea power England, in its war against the land power France, utilized the telluric Spanish partisans, who changed the theater of land war through an irregular space, and later, in World War I, how the land power Germany utilized the submarine as a weapon against the sea power England, thereby opening an unexpected space unknown to the former space of sea war. The former lords of the surface of the sea immediately sought to have the new type of war declared to be irregular, criminal, and even a type of warfare typical of pirates. Today, in the age of submarines with Polaris missiles, both Napoleon's denunciation of the Spanish guerrillas and England's denunciation of German submarines appear to be on one and the same intellectual plane, namely, denunciation of worthless judgments with respect to incalculable spatial changes.

Destruction of Social Structures

The French in Indochina (1946–56) experienced a powerful example of destruction of social structures when their colonial empire collapsed. We have referred to Ho Chi-minh's organization of partisan warfare in Vietnam and Laos, where the communists also utilized the unpolitical civilian population in their struggle. They even corralled the domestic employees of French officers and officials, as well as French army maintenance personnel. They extorted taxes from the civilian population, and perpetrated all types of terrorist acts in order to cause the French to initiate acts of counter-terror against the indigenous population, which incited even more hatred against the French. In short, the modern form of revolutionary war led to many new, sub-conventional means and methods, but a detailed discussion of these is beyond the scope of our discussion.

A commonwealth exists as *res publica*, as a public sphere, and is challenged if a non-public space develops within it, which actually repudiates this public sphere. Perhaps this explanation is sufficient to demonstrate that the partisan, who displaced the technical-military consciousness of the 19th century, suddenly reappeared as the focus of a new type of war, whose meaning and goal was destruction of the existing social order. This becomes obvious in the changed practice of hostage-taking. In the Franco-German War, German troops took dignitaries of an area as hostages for protection against the *Francs-tireurs*: mayors, pastors, doctors, and notaries. Respect for such dignitaries and notables could be used to pressure the whole population, because social respect for such typically bourgeois classes was very strong. Precisely these bourgeois classes became the real enemy in the revolutionary civil war of communism. Given the situation, whoever used such dignitaries as hostages worked for the communist side. The communists were able to use this type of hostage-taking so purposefully that, if necessary, they could initiate it either to destroy a particular bourgeois class or to force it to the communist side.

In Schroers' book on the partisan, this new reality is well-recognized. As it reports, in partisan warfare a truly effective hostage-taking is possible only against the partisans themselves or against their closest operatives. Otherwise, one only creates new partisans. Conversely, to the partisan, every soldier of the regular army, every man in uniform, is a hostage. As Schroers writes: "Every uniform should feel threatened, and thereby everything that it stands for."[82]

One need only to think this logic of terror and counter-terror through to the end, and then to apply it to every type of civil war in order to see the destruction of social structures at work today. A few terrorists are able to threaten great masses. Wider spaces of insecurity, fear, and general mistrust are added to the narrower space of open terror, creating a "landscape of treason," as Margret Boveri has described in a series of four remarkable books.[83]

82. Schroers, *Der Partisan: Ein Beitrag zur politischen Anthropologie*, op. cit., pp. 33f. Formal prohibitions of hostage-taking (as in Art. 34 of the Fourth Geneva Convention) do not come to grips with modern methods of effective hostage-taking of whole groups; cf. p. 72.

83. Margret Boveri, *Treason in the Twentieth Century*, translated from the German by Jonathan Steinberg (London: Macdonald, 1961). The subjects of these books are not only partisans. But the "abysmal confusion" of a "*landscape of treason*" permits all boundaries of legality and legitimacy to become "hopelessly blurred," so that the drive to obtain a general form of the partisan becomes obvious. I have demonstrated this with the example of Rousseau in "Dem wahren Johann Jakob Rousseau," in *Zürcher Woche*, No. 26 (June 28–29, 1962), cf. notes 20, 21, and 23. From this "abysmal confusion," Armin Mohler, as a historian, concludes that one "tentatively reaches the multiform figure of the partisan...only with historical description. From a greater distance, that sometimes may look different. From a still longer view, every attempt at an intellectual or poetic mastery of this landscape produces only enigmatic, highly significant fragments that are symptomatic of the

Most nations of the European continent have experienced this new reality both physically and personally during the course of two world wars and postwar situations.

The Global-Political Context

Also our third aspect, entanglement in global political fronts and affairs, has long been known to the public at large. The autochthonous defenders of the homeland, who are willing to die *pro aris et focis* [for altar and hearth], the national and patriotic heroes who take to the woods, i.e., all that is characteristic of the elemental, telluric power to repel a foreign invader, in the meantime has come under international and supranational control, which helps and supports, but only in its own interests, i.e., completely other-directed, globally aggressive goals, which it either protects or abandons. The partisan then ceases to be essentially defensive. He becomes a manipulable tool of global revolutionary aggressivity. He simply becomes fired up, and is deceived about why he undertook the struggle and about the roots of its telluric character and the legitimacy of his partisan irregularity.

In some way, as an irregular fighter, the partisan always depends on assistance from a regular power. This aspect of the matter is known and recognized. The Spanish guerrilla found his legitimacy in his defensive character and in his accord with a kingdom and a nation; he defended the soil of the homeland against a foreign invader. But Wellington also participated in the Spanish guerrilla war, and the struggle against Napoleon was pursued with English help. Voller Ingrimm often reminded Napoleon that England was the real instigator and also the real beneficiary

time," as stated in a review of Schroers' book in *Das Historisch-Politische Buch*, Vol. 8 (1962). Mohler's conclusion and the judgment implicit in it naturally also concerns our own attempt at a theory of the partisan, and we will keep that in mind. Our attempt also would be void and pointless if our categories and concepts were as little representative of the facts as those which until now have been expressed to refute and exclude the concept of the political.

of the Spanish partisan war. Today, the context is much more to the fore, owing to the continuous intensification of technical means, which make the partisan dependent on the constant help of a community that is in a technical-industrial position to provide him with the newest weapons and machines.

If several interested third parties become involved concurrently, then the partisan has space for his own politics. That was Tito's situation in the last years of World War II. In the partisan struggles in Vietnam and Laos, the situation was complicated by the fact that the struggle between Soviet and Chinese communism had become acute. With the support of Peking, more partisans have been infiltrated into North Vietnam through Laos; in effect, that is a stronger help for the Vietnamese communists than the support of Moscow. The leader of the war of liberation against France, Ho Chi-minh, was a follower of Moscow. The decisive factor in the decision whether to side with Moscow or Peking, or to seek other alternatives, depending on the situation, became the amount of assistance.

For such highly political relations, Schroers' book on the partisan found an apt formula; it speaks of the *interested third party*. That is a good expression. This interested third party is here not some banal figure, as is usually meant by this amusing proverbial expression. Rather, he is essential to the situation of the partisan, and thus also to his theory. The powerful third party not only provides weapons and munitions, money, material support, and all types of medicine, he also creates the type of political recognition that the irregular fighter needs in order not to be considered in the unpolitical sense of a thief or a pirate, which here means: not to sink into the criminal realm. From a longer perspective, the irregular fighter must be legitimated by the regular, and this means two possibilities are open to him: recognition by an existing regular power, or achievement of a new regularity through his own power. That is a difficult alternative.

To the extent that the partisan is motorized, he leaves his own turf and becomes more dependent on technical-industrial means, which he needs for his struggle. This also causes the power of the interested third party to grow, so that it ultimately reaches planetary proportions. Thus, all these aspects of the contemporary partisan appear to flow into the technical-industrial aspect that dominates everything.

The Technical-Industrial Aspect

The partisan also participates in the development, in the progress of modern technology and its science. The old partisan, who wanted to take his pitchfork in hand after the 1813 Prussian *Landsturm* edict, today appears comical. The modern partisan fights with automatic pistols, hand grenades, plastic bombs, and perhaps soon also with tactical atomic weapons. He is motorized, and linked to an information network with clandestine transmitters and radar gadgetry. With airplanes, he is supplied weapons and food from the air. However, he is also, as today (1962) in Vietnam, combatted with helicopters and starvation. Both he and his enemies keep step with the rapid development of modern technology and its type of science.

An English maritime expert called piracy the "pre-scientific stage" of sea war. In the same spirit, he would define the partisan as the pre-scientific stage of land war, and declare this to be the only scientific definition. Yet, this scientific definition is also and again immediately surpassed, because the disparity between sea war and land war is itself caught in the vortex of technical-industrial progress, and today appears to the technician to be pre-scientific, thus already played out. The dead ride fast, and if they become motorized, they move even faster. In any case, the partisan, on whose telluric character we have focused, becomes the irritant for every purpose-rational and value-rational thinking person. He provokes nothing short of a technocratic affect. The paradox of his existence exposes an incongruity: the industrial-technical perfection of the equipment of a modern regular army *vis-à-vis* the pre-industrial agrarian primitivism of real fighting partisans. Precisely this incongruity provoked Napoleon's rage against the Spanish guerrillas, and, with the progressive development of industrial technology, this incongruity has been intensified still more.

As long as the partisan was only a "light troop," i.e., a tactically-mobile hussar or sharpshooter, his theory remained a speciality of the science of war. Only revolutionary war made him a key figure of world history. However, what is one to make of him in the age of atomic weapons of mass destruction? In a thoroughly-organized technical world, the old, feudal-agrarian forms and concepts of combat, war, and enmity disappear. That is obvious. But do combat, war, and enmity thereby also disappear and become nothing more than harmless social conflicts? When the internal, immanent rationality and regularity of the thoroughly-organized technological world has been achieved in optimistic opinion, the partisan becomes perhaps nothing more than an irritant. Then, he disappears simply of his own accord in the smooth-running fulfillment of technical-functional forces, just as a dog disappears on the freeway. In a technologically-focussed fantasy, he is neither a philosophical, a moral, nor a juridical problem, and hardly one for the traffic cop.

That certainly would be a purely technical view of the technical-optimistic aspect. It would anticipate a new world with a new man. As is well-known, the old Christianity and, two millennia later, in the 19th century, the new Christianity (socialism) had such expectations. Both lacked the totally-destructive efficiency of modern technical means. But pure technicity, as obtains in such purely technical reflections, produces no theory of the partisan, but only an optimistic or pessimistic series of more or less powerful assertions of value and non-value. Value, as Ernst

Forsthoff aptly put it, has "its own logic," namely, the logic of non-value and the destruction of the bearer of non-value.[84]

The prognosticator of widespread technical optimism cannot avoid the answer to what concerns him, namely, self-evident assertions of value and non-value. He believes that an irresistible industrial-technical development of mankind in and of itself transfers all problems, all former questions and answers, all former types and situations to a completely new level, on which the old questions, types, and situations become practically unimportant, just as questions, types, and situations of the Stone Age did after the transition to a higher culture. Then, the partisan dies out, as did the Stone Age hunter, to the extent that he is not equipped to survive and to assimilate. At any rate, he becomes harmless and unimportant.

But what if this human type that until now has produced the partisan succeeds in adapting to the technical-industrial environment, avails himself of the new means, and becomes a new type

84. See his famous article, "Die Umbildung des Verfassungsgesetzes" (1959). The asserter of value asserts with his value *eo ipso* always a non-value; the meaning of the assertion of non-value is the destruction of the non-value. This simple state of affairs is demonstrated not only in the praxis evident in the 1920 publication *Die Vernichtung des lebensunwerten Lebens* [The Destruction of Life Lacking Any Value] (although this example alone should suffice); it revealed itself at the same time and with the same naive, unsuspecting demeanor also in the theoretical stance taken by Heinrich Rickert (*System der Philosophie*, Vol. I, 1921, p. 17): there is no negative existence, only negative values; the urge to negation is the criterion for this, i.e., that something is part of the sphere of values; negation is the essential act of asserting values. Cf. my 1961 article, "Die Tyrannei der Werte," published in *Revista de Estudios Politicos*, No. 115 (1961), pp. 65–81 [republished in Carl Schmitt, Eberhard Jüngel, and Sepp Schelz, *Die Tyrannei der Werte*, ed. Sepp Schelz (Hamburg: Lutherisches Verlagshaus, 1979), pp. 9–43], and the article titled "Der Gegensatz von Gesellschaft und Gemeinschaft, als Beispiel einer zweigliedrigen Unterscheidung. Betrachtungen zur Struktur und zum Schicksal solcher Antitheses," in the *Festschrift* for Prof. Luis Legaz y Lacambra (Santiago de Compostela: 1960), Vol. I, pp. 174ff.

of partisan? Can we then say that a technical-industrial partisan has developed? Is there any guarantee that the modern means of mass destruction always will fall into the right hands, and that an irregular struggle is inconceivable? Compared with any progress-optimism, progress-pessimism and its technical fantasies still have a greater field than most people are aware of today. In the shadow of the contemporary atomic balance of the world powers, under the glass cover, so to speak, of their enormous means of mass destruction, a margin of limited and bracketed war with conventional weapons and even means of mass destruction could be outlined, i.e., a situation that the world powers either could be mutually open to or silent about. That would provide one of these world powers with a controlled war, and would be something like a dogfight.[85] It would be the apparently harmless play of a precisely-controlled irregularity and an "ideal disorder," ideal insofar as it could be manipulated by the world powers.

In addition, there is also a radical-pessimistic *tabula-rasa* solution to the technical-industrial fantasy. In an area devastated by modern weapons of mass destruction, of course everybody

85. "Finally, together with the totality of war, certain methods of reaching a compromise and of avoiding an unmanageable arms race also and always develop simultaneously. Then, each side seeks to avoid total war, which accordingly brings with it total risk. Thus, in the postwar period, the so-called military reprisals (the Corfu Conflict, 1923, and Japan-China, 1932), then the attempts at non-military economic sanctions according to Art. 16 of the League of Nations Charter (autumn 1935 against Italy), and finally, certain methods of testing strength on foreign soil (Spain 1936/37) in a way found their true meaning only in the narrowest context of the total character of modern war. They are transitional, intermediate arrangements between total war and real peace; they obtain their meaning in that total war remains in the background as a possibility, and an understandable caution is manifest in the closing of certain gaps. Only from this perspective can one also understand the science of international law." See my article, "Totaler Feind, totaler Krieg, totaler Staat," in Carl Schmitt, *Positionen und Begriffe im Kampf mit Weimar—Genf—Versailles 1923–1939* [1940] (Berlin: Duncker & Humblot, 1988), p. 236.

would die—friends and enemies, regular and irregular fighters. Nevertheless, it is technically conceivable that a few men might survive the night of bombs and rockets. In view of this eventuality, it would be practically and even rationally necessary to plan jointly the post-bomb situation, and today there are men trained to enter the bombed-out area and immediately to occupy the bomb crater and the destroyed area. Then, a new type of partisan of world history could add a new chapter with a new type of space-appropriation.

Thus, our problem is widened to planetary dimensions. It even reaches still further to supra-planetary dimensions. Technical-industrial progress makes possible the journey into cosmic spaces, and thereby opens up equally immeasurable new possibilities for political conquests, because the new spaces can and must be appropriated by men. Old-style land- and sea-appropriations, as known in the previous history of mankind, will be followed by new-style space-appropriations. *Appropriation* will be followed by *distribution* and *production*. In this respect, despite all further progress, things will remain as before. Technical-industrial progress will create only a new intensity of new appropriations, distributions, and productions, and thereby only intensify the old questions.

Given the contemporary antithesis of East and West, and especially the gigantic race for the immeasurably great new spaces, it is above all a question of political power on our planet, which appears to have become so small. Only he who ostensibly dominates an earth that has become so tiny will be able to appropriate and to utilize new spaces. Consequently, these immeasurable spaces also become potential battlefields, and the domination of this earth hangs in the balance. The famous astronauts and cosmonauts, who formerly were only propaganda stars of the mass media (press, radio, and television), will have the opportunity to become cosmopirates, even perhaps to morph into cosmopartisans.

Legality and Legitimacy

In the development of the partisan, the figure of General Salan enters the picture as an illuminating, symptomatic example of the last stage. In him, the experiences and consequences of regular army war, colonial war, civil war, and partisan war meet and intersect. Salan thought all these experiences through to the end in the compulsory logic of the old adage: with a partisan, one fights like a partisan. He did so consistently, not only with the courage of a soldier, but also with the precision of a General Staff officer and the exactitude of a technocrat. The result was that he was transformed into a partisan, and ultimately declared civil war on his own commander-in-chief and government.

What is the innermost core of such a fate? Salan's chief defense counsel, Maître Tixier-Vignancourt, found a formula in his formidable closing argument of May 23, 1962, that contains the answer to our question. Regarding Salan's activity as head of the OAS, he said: "I must observe that, instead of being a great military leader at the head of an organization, an old militant communist would have chosen a different course of action than did General Salan."[86] That is the decisive point: a professional revolutionary would have acted differently. He would have taken a different position than did Salan not only on the interested third party, but in general.

The development of the theory of the partisan from Clausewitz, over Lenin, to Mao has been pushed forward by the dialectic of regular and irregular fighters, professional officers and professional revolutionaries. Given the doctrine of psychological warfare, which French officers in Indochina took over from Mao, the development is not some type of *ricorso*, going back to the beginning and origin. Here, there is no return to the beginning. The partisan can wear the uniform, and can transform himself into a good, regular fighter, even into an especially heroic regular

86. *Le Procès de Raoul Salan*, op. cit., p. 530.

fighter, similar perhaps to what one says of a poacher, i.e., that he would make an especially capable forest ranger. Yet, that is all too abstract. But the absorption of Mao's theory by those French professional officers in fact had something abstract about it and, as was said once in Salan's trial, had something of a *esprit géometrique* [geometric spirit].

The partisan easily can transform himself into a good soldier in uniform; by contrast, for a good professional officer, the uniform is more than a costume. The regular soldier can become an institutionalized professional; the irregular fighter cannot. The professional officer can become a member of a monastic order, as did Saint Ignatius Loyola. The transformation into pre- and sub-conventional orders means something different. With technocratic expertise, one can vanish in the dark, but it is impossible with such expertise to transform the dark into a sphere of conflict that would culminate in the destruction of the former theater of empire and the dislocation of the great stage of the official public sphere. The Acheron cannot be calculated, and does not follow any magic formula, be it of a still very intelligent mind, be it found in a still very desperate situation.

It is not our task to second-guess the intelligent and experienced officers of the Algerian putsch of April 1961 and the organizers of the OAS regarding what for them were very obvious and concrete questions, in particular regarding either the impact of acts of terror against a civilized European population or regarding the *interested third party*. This last question is meaningful enough. We have mentioned that the partisan needs legitimation if he is to be included in the political sphere and not simply to sink into the criminal realm. The question is not resolved with common antitheses of legality and legitimacy that have become standard today, because in this case legality proves to have the much stronger validity, yes, even the proper validity that it originally had for a proper republican, namely, as the rational, the progressive, the only modern, in a word, the *highest* form of legitimacy.

I do not want to repeat what I have been saying for more than thirty years about this theme. But a few remarks are needed to understand the situation of republican General Salan in the years 1958–61. The French Republic is a government ruled by law; that is its basis, i.e., that it will not be destroyed by the antithesis of right and law or by the distinction of right as a higher authority. Neither the judiciary nor the army is above the law. There is a republican legality, and that is the only form of legitimacy in the Republic. For a true republican, everything else is a sophism hostile to the Republic. The public prosecutor in Salan's trial thus took a simple and clear position; he always and again took refuge in the "sovereignty of the law," which remained superior to every other authority or norm. If this were not so, there would have been no sovereignty of the law. But there was, and it transformed the irregularity of the partisan into a deadly illegality.

In this respect, Salan had no other argument than his statement that, on May 15, 1958, he had supported General de Gaulle in taking power from the legal government, that he then, before God, had committed his conscience, his pairs [regular and irregular, legal and illegal], and his fatherland to the struggle, and that now, in 1962, he saw that he had been deceived and betrayed about what, in May 1958, had been declared to be sacrosanct and what had been promised.[87] He appealed to the nation against the state, to a higher type of legitimacy against legality. Earlier, General de Gaulle often had spoken of traditional and national legitimacy, and had opposed it to republican legality. That changed in May 1958. Also, the fact that General de Gaulle's own legality had obtained only since the referendum of September 1958 did not alter the fact that he had republican legality on his side. Salan was

87. Ibid., p. 85.

compelled to realize that, for a soldier taking a desperate position, he had to fight regularity with irregularity and to transform a regular army into a partisan organization.

Yet, in and of itself irregularity does not amount to anything. It is simply illegality. To be sure, it is indisputable that today there is a crisis of law and thus of legality. The classical concept of law, whose value alone is to sustain republican legality, is being challenged in theory and in practice. In Germany, the appeal to right as opposed to law has become axiomatic to jurists and scarcely is noticed. Today, non-jurists simply say legitimate (not legal) when they want to prove that they have right on their side. Salan's case demonstrates that even the legality that is challenged in the modern state is stronger than any other type of right. That is a manifestation of the decisionistic power of the state and its transformation of right into law. Here, we need not elaborate on this.[88] Perhaps that all will change when the state "withers away." In the meantime, legality is the irresistible functional mode of every modern state army. The legal government decides who is the enemy against whom the army must fight. Whoever claims the right to determine the enemy also claims the right to his own new legality, if he refuses to recognize the enemy determined by the former legal government.

88. The Jacobins of the French Revolution still were conscious of the sanctity of their concept of law; they were politically intelligent and courageous enough to distinguish sharply between *loi* and *mesure*, law and measure [meaning expedient action], to say openly that measure is *revolutionary*, and to reject any obliteration of the difference between law and measure as with such conceptual montages as measure-law. Unfortunately, this origin of the republican concept of law is not recognized by Karl Zeidler in *Massnahmegesetz und "klassisches" Gesetz: Eine Kritik* (Karlsruhe: C. F. Müller Verlag, 1961); thus, he misses the essential problem. Cf. gloss 3 to my 1932 monograph, "Legalität und Legitimität," in *Verfassungsrechtliche Aufsätze*, op. cit., p. 347, and the references to legality and legitimacy in the index, pp. 512–13. Roman Schnur has written a more substantial work, not yet published, titled *Studien zum Begriff des Gesetzes*.

The Real Enemy

A declaration of war is always a declaration of an enemy. That is obvious, as is a declaration of civil war. When Salan declared civil war, in effect he made two enemy declarations: one, to continue regular and irregular war on the Algerian front; the other, to open an illegal and irregular civil war against the French government. Nothing makes the inescapability of Salan's situation so clear as this declaration of two enemies. Every two-front war raises the question of who then is the real enemy. Is it not a sign of inner conflict to have more than one real enemy? If the enemy defines us,[89] and if our identity is unambiguous, then where does the doubling of the enemy come from? An enemy is not someone who, for some reason or other, must be eliminated and destroyed because he has no value. The enemy is on the same level as am I. For this reason, I must fight him to the same extent and within the same bounds as he fights me, in order to be consistent with the definition of the real enemy by which he defines me.

Salan considered the Algerian partisans to be the absolute enemy. Suddenly, a much more serious, more intensive enemy appeared at his rear—his own government, his own commander-in-chief, his own comrades. Suddenly, he recognized a new enemy in his comrades of yesterday. That is the core of Salan's case. The

89. [Tr. Schmitt's statement in German reads: "Der Feind is unsre eigene Frage als Gestalt," which literally means that the enemy is the shape or configuration of our own question. In his postwar notebooks, Schmitt is more specific about the meaning of this phrase. He writes: "*Historia in nuce* [history in a nutshell]. Friend and Enemy. The friend is he who affirms and confirms me. The enemy is he who challenges me (Nuremburg 1947). Who can challenge me? Basically, only myself. The enemy is he who defines me. That means *in concreto*: only my brother can challenge me and only my brother can be my enemy." See Carl Schmitt, *Glossarium: Aufzeichnungen der Jahre 1947–1951*, ed. Eberhard Freiherr von Medem (Berlin: Duncker & Humblot, 1991), dated February 13, 1949, p. 217. Also indicative is Schmitt's statement: "Tell me who your enemy is, and I will tell you who you are. Hobbes and the Roman Church: the enemy is he who defines me." Ibid., dated May 23, 1949, p. 243.

comrades of yesterday were revealed to be the more dangerous enemy. There must be a confusion in the concept of the enemy, which is related to the theory of war. We now turn to a clarification of this confusion at the end of our investigation.

A historian would find examples and parallels in world history for all historical situations. We have indicated parallels with incidents from the years 1812–13 of Prussian history. We also have indicated how the partisan obtained his philosophical legitimation in the ideas and plans of Prussian military reform from 1808–13 and his historical accreditation in the Prussian *Landsturm* edict of April 1813. Thus, it is not so strange as at first it might appear to be if we utilize the situation of Prussian General Hans von Yorck in the winter of 1812–13 as a counter-example to better elucidate our core question. At first, what comes to mind are the enormous contradictions: Salan, a Frenchman of Left-Republican extraction and modern-technical stamp, as compared with a royal Prussian army general in 1812, who certainly never would have thought of declaring civil war against his king and commander-in-chief. In view of such disparities of time and type, it appears immaterial and even incidental that Yorck also fought as an officer in colonial East India. However, precisely the striking antitheses make it more evident that the core question is the same, because in both cases there was the need to decide who was the real enemy.

Decisionistic exactitude dominates the functioning of every modern organization, especially every modern, regular army. Thus, for the situation of a contemporary general, the core question poses a very precise question as an absolute either-or. The critical alternative of legality and legitimacy is only a result of the French Revolution and its conflict with the restored legitimate monarchy of 1815. In a pre-revolutionary legitimate monarchy, such as the kingdom of Prussia at that time, feudal elements still were maintained, such as the relation between leader and subordinate. Nevertheless, the situation was not "irrational," and had not yet dissolved into a mere calculable functionalism. Already then, Prussia was a very conspicuous *state*, its army could not deny its origin during the rule of Frederick the Great, and it could not be denied that the Prussian military reformer in no sense wanted to return to some type of feudalism, but rather wanted to modernize the army. Nevertheless, to the contemporary observer, the ambiance of the legitimate Prussian monarchy would appear to have been less sharp and cutting, less decisionistically political. But there is no need here to argue with that. There is only the need to be careful that impressions of various times not obscure the core question, namely the question of the real enemy.

In 1812, Yorck commanded the Prussian division that was allied with Napoleon and belonged to the army of the French General Etienne-Jacques Macdonald. In December 1812, Yorck defected to the enemy, to the Russians, and concluded with Russian General Johann von Diebitsch the Treaty of Tauroggen. In the negotiations and the settlement, Lt. Col. von Clausewitz acted for the Russian side as a mediator. The letter dated January 3, 1813, that Yorck addressed to his king and his commanding officer became a famous historical document. With good reason. The Prussian general wrote very respectfully that he had expected from the king a decision as to whether he (Yorck) should move "against the real enemy" or whether the king had condemned the action of his general. In both cases, he sees the same conscientious resignation as in the case of the decision "to await the bullet on the sand pile even as on the battlefield."

The term "real enemy" is worthy of Clausewitz. It hits upon the core of the matter, and is a matter of fact in General Yorck's letter to his king. The fact that the general was prepared "to receive the bullet on the sand pile" accords with the duty of a soldier who was ready to die for his actions, no different from General Salan, who was ready in the graves of Vincennes to shout "*Viva la France*" in front of the execution commander. Yorck's letter obtains its true, tragic, and rebellious significance from the fact that, despite his respect for the king, he reserved the decision regarding who is the "real enemy." Yorck was no partisan, and it never would have occurred to him to become one. Yet, from the standpoint of the meaning and the concept of the real enemy, his becoming a partisan would have been neither senseless nor illogical.

Obviously, this is only a heuristic fiction, which we will entertain for the moment, namely that a Prussian officer would consider elevating the partisan to an idea, if only at the turning point precipitated by the *Landsturm* edict of April 13, 1813. Already a few months later, the idea that a Prussian officer would become a partisan, even as a heuristic fiction, had become grotesque and absurd, and that remained true for as long as there was a Prussian army. How was it possible that the partisan, who in the 17th century had sunk to the level of a *Pícaro*, and in the 18th century was a light troop, for a moment, at the turn of the century (1812–13), appeared to be a heroic figure, and then, in our time, over 100 years later, even became a key figure of world history?

The answer is that the irregularity of the partisan remains dependent on the meaning and content of a concrete regularity. After the dissolution that was characteristic of Germany in the 17th century, a regularity developed in the 18th century with cabinet wars, which bracketed war so strongly that it could be conceived of as play in which the light movable troops joined in the game as irregulars and the enemy became merely a conventional enemy, an adversary in a war game. The Spanish guerrilla war began when Napoleon, in autumn 1808, defeated the regular Spanish army. By comparison, when the Prussian regular army was defeated in 1806–7 it immediately concluded a humiliating peace, whereas the Spanish partisans renewed the war in earnest, and against Napoleon no less. The old European continental states, which had become accustomed to convention and play, were put on the defensive. Old regularity no longer was any match for the new, revolutionary, Napoleonic regularity. The enemy again became the real enemy, and war again became real war. The partisan, who defended the national soil against the foreign invader, became the hero, who actually fought against a real enemy. This was the great process that led Clausewitz to his theory and to his book *On War*. Then, 100 years later, the professional revolutionary Lenin blindly destroyed all traditional bracketing of war in his theory of war. War became absolute war, and the partisan became the bearer of absolute enmity against an absolute enemy.

From the Real Enemy to the Absolute Enemy[90]

In the theory of war, it is always the distinction of enmity that gives war its meaning and character. Every attempt to bracket or limit war must have in view that, in relation to the concept of war, enmity is the primary concept, and that the distinction among different types of war presupposes a distinction among different types of enmity. Otherwise, all efforts to bracket or limit war are mere play that do not stand in the way of the outbreak of real enmity. After the Napoleonic wars, irregular war was dislodged from the general consciousness of European theologians, philosophers, and jurists. In fact, there were friends of peace, who saw the end of war altogether in the abolition and proscription of conventional war in the 1907 Hague Convention; and there were jurists who held that every theory of just war is *eo ipso* just,

90. [Tr. The German language makes no distinction between enemy (*Feind*), i.e., a legitimate opponent, whom one fights according to recognized rules and whom one does not discriminate against as a criminal, and a foe, i.e., a lawless opponent, whom one must fight to the death and destroy. For this reason, Schmitt was forced to distinguish between the "real enemy" and the "absolute enemy." For reference, see G. L. Ulmen, "Return of the Foe," and George Schwab, "Enemy or Foe: A Conflict of Modern Politics," in *Telos* (Summer 1987), No. 72, pp. 187–93 and 194–201, respectively.]

as taught by Thomas Aquinas. Nobody had an inkling of what the unleashing of irregular war meant. Nobody had thought of the impact that a victory of the civilian over the soldier would have, when one day the citizen would put on the uniform, while the partisan takes it off in order to fight without one.

Precisely this lack of concrete thinking culminated in the destructive work of professional revolutionaries. That was a great misfortune because, with the bracketing of war, European humanity had achieved something extraordinary: renunciation of the criminalization of the opponent, i.e., the relativization of enmity, the negation of absolute enmity. That really was extraordinary, even an incredibly human accomplishment, that men disclaimed a discrimination and denigration of the enemy.

Even that appears to be in doubt as a result of the partisan. The most extreme intensity is part of the partisan's criteria of political engagement. When Che Guevara said "The partisan is the Jesuit of war," he was thinking of the unconditional nature of political engagement. The personal history of every famous partisan, from Empecinado on, affirms this fact. In enmity, a person who has lost his right seeks to regain it; in enmity, he finds the meaning of his cause and the meaning of right when the framework of protection and obedience within which he formerly lived breaks up, or when the web of legal norms that he formerly could expect of law and a sense of justice is torn apart. Then, the conventional play ceases. Yet, the dissolution of this legal protection still need not lead to the partisan. Michael Kohlhaas,[91] who gave a sense of justice to robbers and murderers, was

91. [Tr. *Michael Kohlhaas* is the title of a novel by Heinrich von Kleist. It is based on Hans Kohlhase, a merchant who lived in Cologne. In October 1532, so the story goes, Kohlhase, while on his way to a fair in Leipzig, was attacked and his horses were taken by the servants of a Saxon nobleman, Günter von Zaschwitz. Having thus been delayed in getting to the fair, Kohlhase suffered a loss of business, and on his return refused to pay Zaschwitz the small sum demanded for return of his horses. Instead, Kohlhase demanded a large

no partisan, because he was not political. He did not fight against a foreign invader or for a revolutionary cause, but rather for his own offended private right. In such cases, irregularity is unpolitical and becomes purely criminal, because it loses the positive connection with regularity. For this reason, the partisan distinguishes himself from the noble or ignoble robber baron.

We have emphasized with respect to the global political context that the *interested third party* has an essential function in that he provides the relation to regularity that the irregularity of the partisan needs to remain in the political sphere. The core of the political is not enmity *per se*, but the distinction of friend and enemy, and the presupposition of friend *and* enemy. The powerful third party interested in the partisan still is able to think and to act very egoistically; his political interest is on the side of the partisan. In effect, that is political friendship, and is a type of political recognition, even if it does not come to public or formal recognition of the party or government fighting a war. The Empecinado was recognized as a political force by his people, by the regular army, and by the English world power. He was no Michael Kohlhaas, and also no Schinderhannes,[92] whose

compensation. Failing in this, he was forced to pay Zaschwitz to recover his horses, but he reserved the right to take further action. Then, unable to obtain redress in a court of law, he challenged not only Zaschwitz, but the whole of Saxony. Acts of lawlessness soon were attributed to him. After attempts made to settle the feud failed, the elector of Saxony, John Frederick I, set a price on Kohlhase's head. Seeking revenge, Kohlhase gathered around him a band of criminals, and began spreading terror throughout the whole of Saxony: travellers were robbed, villages were burned, and towns were plundered. For a time, the authorities were powerless to stop the outrages, but in March 1540 Kohlhase and his principal associate, Georg Nagelschmidt, were seized, and soon were broken on the wheel in Berlin.]

92. [Tr. Hans Bückler, the Schinderhannes [oppressor], became notorious for highway robbery, attacks on farmsteads, and blackmail. He called himself "Johannes of the woods," and used this name to sign passes for unrestricted movement. As "King of Soon Forest," another name Bückler assumed, he appropriated the right to issue these passes as well as to collect taxes. In 1801,

interested third parties were fences, i.e., receivers of stolen goods. By contrast, Salan's political situation foundered in a desperate tragedy, because he became illegal internally, in his own fatherland, whereas externally, in world politics, not only did he not find an interested third party, but, on the contrary, he ran up against the united enemy front of anticolonialism.

The partisan has a real, but not an absolute enemy. That follows from his political character. Another limitation of enmity follows from the telluric character of the partisan. He defends a piece of land with which he has an autochthonous relation. His fundamental position remains defensive, despite the intensive mobility of his tactics. He comports himself in the same manner as did Joan of Arc before the clerical court. She was no partisan. She fought a regular war against the English. When the clerical judge asked her the question—a theologically-loaded question—whether she was contending that God hated the English, she answered: "I do not know whether God loves or hates the English; I only know that they must be driven out of France." Every normal partisan would have given this answer in defense of the national soil. Such a fundamentally defensive position also presupposes a fundamental limitation of enmity. The real enemy will not be declared to be an absolute enemy, also not the last enemy of mankind.[93]

farmers began revolting against him. Eventually, he was caught and brought to trial. Bückler, who had enlisted in the Austrian army under a false name, made a complete confession. But he asked for mercy for his wife, who had just given birth to their son, and for some of his followers. This plea so impressed the public in the courtroom that suddenly he was not seen as a notorious robber, but as a gentleman robber baron and helper of the poor, like Robin Hood. The judges in Mainz did not share this opinion. Bückler and twenty of his followers were sentenced to death. Carl Zuckmeyer wrote a play called *Schinderhannes*, which premiered in Berlin in 1927, in which Bückler is portrayed as a literary hero.]

93. "Such a war (which at any given time could be declared to be the last war of mankind) is of necessity unusually intense and inhuman, because, by

Lenin shifted the conceptual center of gravity from war to politics, i.e., to the distinction of friend and enemy. That was significant and, following Clausewitz, a logical continuation of the idea that war is a continuation of politics. But Lenin, as a professional revolutionary of global civil war, went still further and turned the real enemy into an absolute enemy. Clausewitz spoke of absolute war, but always presupposed the regularity of an existing state. He could not conceive of a state becoming an instrument of a party, and of a party that gives orders to the state. With the absolutization of the party, the partisan also became absolute and a bearer of absolute enmity. Today, it is not difficult to see through the intellectual artifice produced by this change of the enemy concept. By comparison, another type of absolutization of the enemy is much more difficult to refute today, because the existing reality of the nuclear age appears to be immanent.

Technical-industrial development has intensified the weapons of men to weapons of pure destruction. For this reason, an infuriating incongruity of protection and obedience has been created: half of mankind has become hostage to the rulers of the other half, who are equipped with atomic weapons of mass destruction. Such absolute weapons of mass destruction require an absolute enemy, and he need not be absolutely inhuman. Certainly, it is not the weapons of mass destruction that destroy, but rather men, who destroy other men with these weapons. Already in the 17th century, the English philosopher

transcending the limits of the political framework, it simultaneously degrades the enemy into moral or other categories and is forced to make of him a monster that not only must be defeated, but also destroyed. In other words, he is an enemy who no longer must be compelled only to retreat into his borders. The feasibility of such a war is particularly illustrative of the fact that war as a real possibility is still present today, and this fact is crucial for the friend-enemy antithesis and for the recognition of politics." See Carl Schmitt, *The Concept of the Political*, tr. George Schwab (Chicago: The University of Chicago Press, 1996), pp. 36–37 (translation altered).

Thomas Hobbes understood the core of the process, and formulated it very precisely,[94] although at that time (1659) the weapons by comparison were still quite harmless. Hobbes said: the man who feels threatened by other men is even more threatened than by an animal. The weapons of men are more dangerous than the so-called natural weapons of animals, for example: teeth, claws, horns, or poison. The German philosopher Hegel added: weapons are the essence of the fighter.

Concretely, this means supra-conventional weapons presuppose supra-conventional men, not only as a postulate of a far-distant future, but as an existing reality. Thus, the ultimate danger exists not even in the present weapons of mass destruction and in a premeditated evil of men, but rather in the inescapability of a moral compulsion. Men who use these weapons against other men feel compelled morally to destroy these other men, i.e., as offerings and objects. They must declare their opponents to be totally criminal and inhuman, to be a total non-value. Otherwise, they are nothing more than criminals and brutes. The logic of value and non-value reaches its full destructive consequence, and creates ever newer, ever deeper discriminations, criminalizations, and devaluations, until all non-valuable life has been destroyed.

In a world in which opponents mutually consign each other to the abyss of total devaluation before they can be destroyed physically, new types of absolute enmity must be created. Enmity becomes so frightful that perhaps one no longer should speak of the enemy or enmity, and both should be outlawed and damned in all their forms before the work of destruction can begin. Then, the destruction will be completely abstract and completely absolute. In general, it no longer would be directed against an enemy, but rather would serve as a given objective realization of the

94. Thomas Hobbes, *Man and Citizen*, a translation of *De Homine*, ed. and intro. Bernard Gert (Atlantic Highlands, NJ: Humanities Press), IX, 3.

highest values, for which no price would be too high. Only the denial of real enmity paves the way for the destructive work of absolute enmity.

In 1914, the nations and governments of Europe stumbled into World War I without real enmity. Real enmity arose only out of the war, when a conventional state war of European international law began, and ended with a global civil war of revolutionary class enmity. Who can prevent unexpected new types of enmity from arising in an analogous, but ever more intensified way, whose fulfillment will produce unexpected new forms of a new partisan?

The theoretician can do no more than verify concepts and call things by name. The theory of the partisan flows into the question of the concept of the political, into the question of the real enemy and of a new *nomos* of the earth.

Also from Telos Press Publishing

Timothy W. Luke and Ben Agger, eds.
A Journal of No Illusions: Telos, *Paul Piccone, and the Americanization of Critical Theory*

Carl Schmitt
Hamlet or Hecuba: The Intrusion of the Time into the Play

Victor Zaslavsky
Class Cleansing: The Massacre at Katyn

Ernst Jünger
On Pain

Paul Piccone
Confronting the Crisis: Writings of Paul Piccone

Matthias Küntzel
Jihad and Jew-Hatred: Islamism, Nazism, and the Roots of 9/11

Carl Schmitt
The Nomos *of the Earth in the International Law of the* Jus Publicum Europaeum

Jean-Claude Paye
Global War on Liberty

Jean Baudrillard
The Mirror of Production

Jean Baudrillard
For a Critique of the Political Economy of the Sign